75-
(DR)

The Galápagos Tortoises: Nomenclatural and Survival Status

PETER C.H. PRITCHARD

The Galápagos Tortoises: Nomenclatural and Survival Status

Peter C.H. Pritchard

Florida Audubon Society,
1331 Palmetto Avenue, Suite 110,
Winter Park, Florida 32789 USA

Chelonian Research Monographs
Number 1

July 1996
Chelonian Research Foundation

***Front Cover*:** The cover picture of *Onan*, a celebrated old male tortoise from Duncan Island (Pinzón), was painted by Naoko Izuhara, an artist resident in Winter Park, Florida. The painting is based upon a color photograph taken by the author in August 1986. The story of *Onan*, a nearly blind but aggressive and spirited animal thought to be well over a century in age, is told by Switak (1995). The name *Onan*, given to the tortoise by Craig MacFarland in the early 1970s, derives from the Biblical *Onan* (Genesis 38:9), who incurred the wrath of God and was slain for spilling his seed upon the ground. The chelonian *Onan* too has been observed attempting to copulate with inanimate objects, including rocks. Sadly but inevitably, *Onan* was found dead within his customary haunts by Gayle Davis in May 1990. His remains were buried on Duncan.

***Frontispiece*:** This painting of buccaneers and Galápagos tortoises by Charles M. Sheldon appeared in *Pioneers in Tropical America* by Sir Harry Johnston (undated, ca. 1912). The vegetation is somewhat fanciful, in that the *Opuntia* and *Jasminocereus* seem to be miniaturized, and *Aloe* and *Melocactus* are not known from the Galápagos. The geographic features are reminiscent of the valley leading up from Tagus Cove towards Volcan Darwin on Albemarle Island (Isabela). However, the tortoise in the foreground, with its saddlebacked shell and upright, thick neck, is a good representation of the tortoise of Abingdon Island (Pinta). The posture, the shell morphology, and even the groove along the underside of the neck (an artifact of the preservation and mounting process, unknown in living tortoises), are virtually identical to one of the two mounted specimens of Abingdon tortoises in the British Museum of Natural History (BMNH 76.6.21.38–39), collected by Commander J. Cookson in 1875 (see Fig. 23).

***Back Cover (Upper)*:** This nineteenth century engraving of unknown provenance portrays whaleboats close to the shore of Albemarle Island (Isabela). The precipitous, raw nature of the terrain resembles Volcan Wolf in the north; but Cerro Azul in the southwest is nearly as steep, and the ruggedness portrayed may have been enhanced by the artist.

***Back Cover (Lower)*:** The posterior end of the bony carapace of a large male tortoise from the Tortoise Reserve on Indefatigable Island (Santa Cruz). The last word, as it were — the end.

Contribution No. 199 of the
Charles Darwin Foundation for the Galápagos Islands

The Galápagos Tortoises: Nomenclatural and Survival Status. By Peter C.H. Pritchard.
Chelonian Research Monographs, *Number 1.* Edited by Anders G.J. Rhodin.

ISSN (monograph series): 1088-7105
ISBN (this volume): 0-9653540-0-8 (hard cover); 0-9653540-1-6 (soft cover)

TABLE OF CONTENTS

EDITORIAL PROLOGUE

I take great pleasure in presenting the following monographic work by Peter C.H. Pritchard. Peter and I have been good friends and professional colleagues since the middle 1970s and during this time we have had the marvelous fortune to experience the magic of the Galápagos Islands together on two occasions. The first of those visits, in 1982, was for me the kind of experience that can only be described as one of life's defining moments.

Touring the archipelago first with a group of ecotourists organized by Florida Audubon Society and led by Peter gave me an overview and a chance to appreciate the full experience of the Galápagos. The incredible and varied topography, the teeming seas, the profusion of bird life, and everywhere the evidence of what nature's magnificence offers in its varied flora and fauna impressed me beyond words. Of greatest interest to me, of course, was the herpetological fauna. The land iguanas, marine iguanas, and lava lizards all held incredible fascination and I spent a long time marveling at their profusion and diversity. But as expected, and without a doubt, it was the giant tortoises that captured my imagination and propelled me into the kind of reverie from which one never truly recovers.

Peter C.H. Pritchard with a large male giant tortoise (140 cm carapace length over the curve) in the Tortoise Reserve southwest of Santa Rosa on Indefatigable Island (Santa Cruz), September 1982. Photo by Anders G.J. Rhodin.

After the formal tour of the islands, Peter and I were able to visit several tortoise areas, where I was able to focus fully on the chelonian experience. It was transcendent to see these living reminders of ancient times, to sit with them in quiet solitude, and to share with them the meditative state of an unhurried existence in an ever more intrusive world.

We first visited the Tortoise Reserve southwest of Santa Rosa on Indefatigable Island (Santa Cruz). Here the abundant huge dome-shelled tortoises live in lush, verdant forests and as I shared their world with them for a short while I felt as if I had been transported in a time capsule back to some Cretaceous landscape. A soft rain fell from time to time, the luxuriant vegetation and tortoise carapaces glistened, and I felt privileged to experience the moment.

We next visited Duncan Island (Pinzón). This island is totally different, with a hot, xeric habitat of sunbaked cacti and sparse brush. The tortoises here are small and saddlebacked, unlike the giants on Indefatigable. And they are hard to find. We spent a long time hiking into the high areas where the native tortoises live, and then searched long and hard before we found a few old animals hiding under rocks or brush. Their ancient wrinkled faces and smooth worn shells spoke of a lifetime of struggle for existence in their harsh environment. As I sat on the high hillside in their presence and gazed out past the dry vegetation and rough landscape to the blue waters of the surrounding ocean below I felt a sense of peace and contentment. This was a magical place and its chelonian inhabitants imparted a sense of timelessness to its stark beauty.

A place we did not visit together, but that I had hoped to, was the caldera and crater rim of Volcan Alcedo on Albemarle Island (Isabela). Here the population of tortoises was the most intact — the most primeval — in the entire archipelago. Difficult to reach from the coast and thereby protected from slaughter by nineteenth century whalers, the population was also isolated from southern Albemarle by the lava flows of Perry Isthmus, which had prevented the influx of feral goats to the region. Recently and ominously, however, goats have succeeded in crossing this barrier and are now seriously threatening the tortoise population through widespread habitat destruction as their numbers multiply astronomically. A campaign to save the tortoises of Alcedo has recently been initiated by the Charles Darwin Foundation. We all hope that their efforts are successful in saving these tortoises so that our children and their descendants will derive as much joy and inspiration from their splendor as we have. Our passion for these magnificent creatures must drive our continuing efforts to better understand and protect them. May our legacy to the future be that we succeeded.

Anders G.J. Rhodin

Tortoises in the Mist

Tortoises in the mist on the crater rim above the caldera of Volcan Alcedo, Albemarle Island (Isabela), July 1986. During dry periods tortoises migrate from the dry caldera floor up to the crater rim where the mist (*garua*) of the cloud-catchment drip zone provides them the moisture they need. See the preceding editorial prologue and the account on pp. 71–73 for further details. Photo by Peter C.H. Pritchard.

Tortoises in the mist
timeless creatures, moisture-kissed
neath shrouded trees and mossy lace
since eons past, still exist.

On crater rim, a magic place
tortoises move, at their pace
from caldera floor below
ponderous steps, gentle grace.

Ringed by fire, lava flows
spared the whalers' deadly blows
survive serene, Alcedo home
enchanted isles, Galápagos.

Anders G.J. Rhodin
10 April 1996

The Darwin Connection

"I have not as yet noticed by far the most remarkable feature in the natural history of this archipelago; it is, that the different islands to a considerable extent are inhabited by a different set of beings. My attention was first called to this fact by the Vice-Governor, Mr. Lawson, declaring that the tortoises differed from the different islands, and that he could with certainty tell from which island any one was brought...

The inhabitants, as I have said, state that they can distinguish the tortoises from the different islands; and that they differ not only in size, but in other characters. Captain Porter has described those from Charles and from the nearest island to it, namely, Hood Island, as having their shells in front thick and turned up like a Spanish saddle, whilst the tortoises from James Island are rounder, blacker, and have a better taste when cooked."

— *Charles Darwin* (1845, *Journal of Researches*)

"I knew Darwin. Nice guy."

— *Sauers* (ca. 1980, *The New Yorker*)

ABSTRACT

Pritchard, Peter C.H. 1996. The Galápagos Tortoises: Nomenclatural and Survival Status. Chelonian Research Monographs 1:1–85.

The history of human attention to Galápagos tortoises is broken down into four phases. These are: the Discovery Phase (1535–1840), during which the islands were discovered and early visitors recorded their obervations, often with considerable attention to the giant tortoises; the Taxonomic Phase (1824–1917), during which specimens of the various island tortoise populations (many lacking collecting data) reached major museums and were described and named; the Rescue Phase (1925–1955), during which zoological visitors concluded that the tortoises were doomed in their native habitat and could only be saved by captive breeding in zoological institutions; and the *In Situ* Phase (1957–present), philosophically a reversal of the Rescue Phase, during which increasing efforts were made to conserve, protect, and procreate the tortoises in the Galápagos Islands themselves. Techniques used during the *In Situ* Phase included environmental education programs for Galápagos residents and settlers, control of feral mammals depredating or competing with the tortoises, protection of wild tortoise nests with rock cairns or transfer of nests to artificial incubation, head-starting of young tortoises through the first years of life, and closed-cycle captive breeding. Furthermore, the establishment of Galápagos National Park gave at least nominal legal protection to tortoise habitat throughout the archipelago.

The scientific nomenclature of Galápagos tortoises is examined and found to have many problems. The recently commonly used name, *Geochelone elephantopus*, is not the oldest available name for the species and should no longer be used. The oldest name for Galápagos tortoises is *Testudo californiana*, and under current ICZN regulations this generally unfamiliar name would need to be utilized unless petition is made for its suppression. In the interim, the oldest available familiar name for all Galápagos tortoises is *Testudo nigra* Quoy and Gaimard, 1824, and its use is hereby advocated as the combination *Geochelone (Chelonoidis) nigra*. The various forms from the different islands are all considered as subspecies of *G. nigra*. It is also necessary to petition the ICZN to suppress the name *Testudo ephippium*, a senior synonym of *Testudo abingdoni* from Abingdon Island, but that has been utilized extensively for the Duncan Island tortoise. On Albemarle Island, the names of the various populations and described subspecies of all but the northernmost volcano are synonymized under *Geochelone nigra vicina*, on the grounds that the described differences are either attributable to environmental differences (especially of rainfall, food availability, and humidity), or do not show geographic correlation but are artifacts of age and sex. Several potential but unnamed subspecies are listed, including those of northeastern Chatham Island, northwestern Indefatigable Island, and the extinct forms that may once have inhabited Barrington and Jervis islands. Particular attention is paid to an analysis of the arguments and conclusions of Garman (1917), this being the last comprehensive analysis of the taxonomy of Galápagos tortoises and one that has never been formally challenged despite many deficiencies, as well as certain remarkable examples of insight or prescience.

The survival status history of each of the tortoise populations from the time of first discovery to the present (or until the time of extinction) is reconstructed to the extent that the written record permits. It is observed that considerable conservation progress has been made in the decades since the *In Situ* Phase started in 1957, but in the last two or three years enormous and possibly insurmountable challenges have arisen in the form of a general breakdown of respect for conservation law in the archipelago.

Key Words. – Reptilia; Testudines; Testudinidae; *Geochelone nigra*; tortoise; turtle; conservation; management; repatriation; head-starting; introduced species; extinction; survival status; nomenclature; taxonomy; distribution; natural history; Ecuador; Galápagos

ACKNOWLEDGMENTS

My heartfelt thanks are offered to the memories of four truly great men who, a generation ago, took time to encourage a young Oxford chemistry undergraduate to travel to the Galápagos Islands and to make observations on their giant tortoises. They were Professor Victor van Straelen, Dr. David Lack, Sir Peter Scott, and Sir Alister Hardy.

For several years my field work in the Galápagos Islands was financed by the Worldwide Fund for Nature (then the World Wildlife Fund, U.S. National Appeal). Although this work was primarily on sea turtles, I utilized the opportunity to visit every island in the archipelago from which tortoises had ever been recorded — and arrived on Abingdon just in time to witness the removal of the last living tortoise from that island. During these visits I received many courtesies from a succession of Charles Darwin Research Station Directors, especially Roger Perry, Peter Kramer, and Craig MacFarland. I would also like to thank several veteran Galápagos scientists, including Daniel Weber, Tjitte de Vries, Michael Harris, and Ole Hamann, for sharing their insights into Galápagos biology — and logistics. A succession of boat captains also, especially Johnny Angermeyer, Fritz Angermeyer, Julian Fitter, Hermogenes Moncayo, José Villacis, and the late "Coca Cola," provided reliable charter service — a crucial factor during field work on waterless and uninhabited islands. Arnaldo Tupiza, José Villa, and "Pescadito" guided me to extremely remote *galapaguera* areas of southern Albemarle, Chatham Island, and Hood Island, respectively.

I thank Miguel Cifuentes for serving as field assistant on my expeditions of 1971 and 1972, and feel pride that he so quickly thereafter became an outstanding Director of the Galápagos National Park.

Linda Cayot very generously shared both her time and unrivaled knowledge of the current survival status of the Galápagos tortoises, critiquing a draft of the manuscript and steering me towards literature that I had overlooked. Hobart Smith and Roger Bour also provided masterful critiques of the manuscript and gave me the benefit of their legendary insights into the Code of Zoological Nomenclature. In all cases, the guidance they offered constituted peer review at its finest, and any quality that may be found in the finished work owes much to these three colleagues.

I thank Jean Lescure of the Muséum National d'Histoire Naturelle (Paris) and A.S. Clarke of the Royal Scottish Museum (Edinburgh) for supplying photographs of crucial type specimens, and for permission to publish them. All other photographs are by myself except for a few by Rollo Beck, Roger Bour, Anders Rhodin, and David Snow that are noted in the figure legends.

My wife, Sibille, traveled with me on most of my Galápagos expeditions, and her presence made me a much happier camper in many ways.

Finally, I thank my friend Anders Rhodin, Director of Chelonian Research Foundation, for editing and publishing this work. It is a rare pleasure to have one's compositions edited by one who not only knows the subject matter intimately but who also has a superlative grasp of the principles of English grammar and syntax.

INTRODUCTION

There is much that is ironic, surprising, or even paradoxical about the Galápagos Islands and their famous gigantic tortoises. The association between the islands and the giant tortoises seems so absolute, and indeed is forever formalized in the very name of the archipelago. Yet the Spanish word *galápago* does not refer exclusively to large chelonians, or even just to terrestrial ones. The Conquistadors, possessed of — or acquiring from the native Americans — a modest inventory of vernacular names for tortoises and turtles (*tortuga, caguama, jicotea, galápago, morrocoyo*) parcelled them out to the new chelonian species as they found them in their New World territories. The giant tortoises, the most remarkable inhabitants of the new Pacific archipelago discovered by the off-course Bishop of Panama in 1535, and in a real sense a first-wave of armored Conquistadors themselves, became *galápago*, but so did a common side-necked pond turtle in Venezuela, for example, that rarely exceeds a length of 30 cm. And there is an even smaller, semi-aquatic turtle in English-speaking Trinidad that, to this day, is called *galap*.

Moreover, the very existence of the largest tortoises in the world on minuscule islands in the vast Pacific Ocean strikes many as odd — creatures that represent the epitome of terrestriality, that cannot dive and barely swim, in oceanic island ecosystems where no freshwater turtle species has ever penetrated. Almost all big land animals need big spaces — elephants in the vast plains of east Africa, or bison on the American prairies, come to mind. What is the link between gigantic tortoises and small islands? How did they get there, crossing vast ocean barriers, when aquatic turtles (apart from the truly marine species) failed? How come the Galápagos, with their spectacular herpetofauna and generally absent terrestrial mammalian fauna, can so evocatively mimic Mesozoic times — the glorious Age of Reptiles — while being (the geologists tell us) so new? How is it that such a unique and slow-growing, slow-breeding life form as a giant tortoise could have evolved, and reached gigantic size, so rapidly? And, while the vulnerability of such huge, slow creatures to mankind is understandable, why should such impregnable behemoths be so easily harmed, or even extirpated, by smaller mammals — even black rats, of all things, when lesser tortoises have coexisted with small mammals in continental ecosystems for many millions of years?

These questions are answerable. Firstly, the link between giant tortoises and small oceanic islands is illusory, or at least an artifact of strictly modern, human-dominated conditions. Huge terrestrial chelonians used to be almost everywhere. They have no special affinity for small, remote islands, but they were able to reach them and can survive on them, and these are the only places that we have left for them. Up to and during the Pleistocene, great tortoises, two, three, four, even five feet in length, have been reported as fossils from all continents except Australia and Antarctica, and from many small to large islands as well (Auffenberg, 1974; Hooijer, 1971; Jimenez Fuentes and Carbajosa Tamargo, 1982).

But the spread of mankind over the face of the globe, more than any other factor, has plunged the giants into ignominious retreat. The opportunistic, ingenious new predator, literally as omnivorous as a pig and much more dangerous, proved too much for tortoises beyond a certain size. The great tortoises took too long to reach maturity, found it too difficult to hide, were too helpless and slow. Their armor — a crucial survival factor in pre-human times — was easily sundered by primitive implements. One huge specimen, found as a sub-fossil in Florida, still enclosed the spear that had killed it and retained the charring of the carapace that showed that it had been cooked upside-down on a open fire (Holman, 1981).

After the unequal struggle had wiped the giants off the face of the continents, they lingered on in exile in their own remote little Edens, the archipelagoes of the Galápagos, the Seychelles, and the Mascarenes, for thousands of years, insular ecosystems untouched by the withering hand of man, indeed unknown to man. But finally these island survivors too were found, and their very abundance and toughness under abusive conditions — "tenacious of life" is how they were always described — proved to be their undoing. Whole islands were stripped of their last tortoise so that those whose life styles kept them at sea for months at a time — whalers, buccaneers, explorers — might be victualled with something better than salt pork, something that could continue to live for many months (if you call that living), unfed and unwatered, deep in the wooden hull of a sailing ship. The tortoises of the Mascarenes are gone — the last one died around 1804, a victim of the same frenzy of exploration, colonization, and destruction that wiped out the innocent solitaire, as well as the more famous dodo (North-Coombes, 1976). The tortoises of the Seychelles disappeared also, with the exception of the single population of the remote island of Aldabra. Only in the Galápagos did some diversity of giant tortoise life survive until mankind finally developed a conscience, perhaps in the nick of time. Or perhaps it *was* too late — recent news from the Galápagos is not good.

The Mesozoic "look" of the Galápagos fauna today is an illusion. The spectacular reptiles of the islands all belong to modern groups. Large ground iguanas are also found in the Bahamas and the Greater Antilles, and even the remarkable spectacle of black iguanas sitting on ocean beaches is a familiar one, not just to Galápagos visitors, but to those who have walked the black beaches of the Pacific coast of Central America. And the snakes and lava lizards of the Galápagos would not be considered noteworthy in a mainland tropical setting. The giant tortoises are remarkable for their size, but are famous only because we — humanity — extirpated them on the continents, so we now have to go to Galápagos to see them.

So the islands are no Lost World in the Conan Doyle sense, but by contrast are surprisingly new. The oceanic crust upon which the archipelago sits may be as much as ten million years of age, but the islands themselves probably emerged no more than 3–5 million years ago (Simkin, 1984). Of course, establishing definitive dates or ages is impossible when new lava flows may completely conceal older rocks, but the physiognomy of the islands suggests that the western islands of Narborough and Albemarle, with their active vulcanism, lofty, well-defined calderas, and extensive fresh lava flows, are the newest, while Hood Island, in the extreme southeast — low-lying, probably uplifted, craterless, eroded — may be the oldest. The picture one gets is that of a shifting archipelago plodding out to sea on heavy-footed, slow steps, with newer islands emerging in the west as the eastern ones finally disappear beneath the waves, as the Nazca Plate upon which all the islands sit creeps in the opposite direction, towards the South American continent.

But the Galápagos were never part of the continent, and the tortoises *must* have reached them by drifting, despite the assumption of Gadow (1901) that "accidental transport or migration are out of the question," or the musings of Garman (1917) that they may have been introduced by man. Perhaps they came from South America, and indeed the Humboldt Current sweeps conveniently up the coast of Chile and Peru before diverting westwards at Equatorial latitudes. But the cis-Andean seaboard, washed by the cold Humboldt current, has probably long been an area of remarkable sterility, and another possibility is that the tortoises arrived from the Antilles during a period when the Central American Isthmus was still a series of separate islands. Certainly, fossil giant tortoises have been found in several of the Antilles, including Cuba, Mona, and Navassa (Williams, 1950, 1952; Auffenberg, 1974).

But wherever they came from, they were almost certainly already very large. Small tortoises frequently reach islands close to continental shorelines, and there they may reach high densities, although re-colonization seems to occur frequently enough for such populations not to differ-

Figure 1. An old male Duncan Island tortoise, *Onan*, surveys his arid terrain. August 1986.

Figure 2. Three adult Duncan Island tortoises in a *galapaguera* on a clifftop near the ocean. March 1970.

Figure 3. The snow-white anterior head coloration of an adult male Hood Island tortoise. September 1982.

Figure 4. Spuriously ferocious appearance of a tortoise from Indefatigable Island. September 1986.

entiate from those of the mainland. But islands beyond a few hundred kilometers seem to be reachable only by giants; and if "stepping-stone" islands are absent, even they can apparently not survive the trip to truly remote, mid-ocean islands or archipelagoes.

Lines of reasoning to bolster the argument that gigantism was a pre-adapted condition, useful for successful colonization of oceanic islands, include the following:

1) The newness of the islands in question argues against there having been time for an *in situ* evolution of gigantism.

2) The island of Aldabra, in the Indian Ocean — the only other island with surviving, indigenous giant tortoises — is thought to have been completely submerged several times during the last few tens of thousands of years (Braithwaite et al., 1973), yet each time, it was recolonized by giant tortoises. They must have arrived by drifting from other Indian Ocean islands; but note that it cannot happen again, because tortoises are now extinct on all the potential "donor" islands.

3) In the ocean environment, a giant tortoise would have many survival advantages over a small one. It would have more extensive fat reserves; it could hold its head higher above the waves, thus avoiding accidental ingestion of sea water; it would have a more favorable surface/volume ratio, thus reducing osmotic water loss; and, once (*Deo volente*) a live landfall were made, a giant would be better equipped for locomotion in broken or boulder-strewn littoral terrain, for survival under uncertain or extreme thermal conditions, or toleration of protracted dessicating conditions.

4) The variation observed in the surviving populations of Galápagos tortoises (Figs. 1–5) suggests that the post-arrival specializations that have occurred — specifically, the development of saddlebacked, flared shell forms, generally on smaller, more arid islands — have been characterized by a *reduction* in adult size (partly genetic, partly a result of erratic availability of food and water). Far from "islands creating giants," the real trend may have been towards secondary dwarfism, although it is possible that the trend was complicated or blurred by sexual selection pushing in the opposite direction, the successful males generally being the biggest animals.

Why are the giant tortoises so vulnerable, so subject to utter disappearance within decades, or at most a century or two, of discovery? They don't seem delicate, and even in relatively mediocre zoos without even a close facsimile of an appropriate climate, where tortoises of other species rarely survive for long, Galápagos tortoises may live, without problems, for many decades — although breeding them is another story.

The answer is complex. Firstly, massiveness and physical impregnability evolve under conditions of protracted stability, and are not the keys to survival or dominance in the fast-changing modern world. Today the fellow travelers on Planet Earth that threaten to dominate us are viruses, not *Velociraptor*; tsetse flies, not *Triceratops*.

Moreover, while life may be hard on islands that are frequently without water, and sometimes scorched by vulcanism as well as the tropical sun, the absence of large predators has made the tortoises "soft" in certain ways, to use an unlikely metaphor. On the continents, tortoises have been forced to survive by concealment — hiding in rocky crevices, burrows, or dense vegetation; by crypsis, with camouflaging shell markings or designs; by elaborate modifications of the armor, with anterior extensions to protect the extremities, or by hinging of various sections of the shell, to allow crucial life functions like mating and oviposition without leaving the animal open to attack through gaping shell openings; by ingrained, appropriate, cautious behavioral patterns that reduce the exposure to discovery; and by equilibrium population densities that are often quite modest. In many continental areas where tortoises are quite successful, they nevertheless are so inconspicuous that they are hardly ever seen.

On the islands, these constraints disappear. The tortoises, instead of being elusive masters of survival, become dominant. They achieve high densities, can fall asleep or feed wherever they wish without any obligation to keep out of sight, and can modify their rigid carapacial envelope

Figure 5. Copulating tortoises in the lush caldera of Volcan Alcedo, Albemarle Island. March 1975.

(towards saddlebacked forms, or reduced thickness) in ways that may facilitate oviposition, thermoregulation, or locomotion in broken terrain, but that certainly do not enhance security from predators. The latter, after all, were absent until the dogs and pigs arrived. So what worked well under virgin conditions becomes a liability under violated ones.

The vulnerability of hatchling Galápagos tortoises to rats is curious. After all, rats are ubiquitous under continental regimes where tortoises have always been exposed to them. Moreover, hatchling and small juvenile tortoises of all species are virtually defenseless, with scant bone in their "shells," and offering a nutritious, lipid-rich package that can neither outrun a predator nor even turn and nip it on the snout.

The explanation probably lies in the fact that the ecological scenario that allows some hatchling tortoises to survive is always a delicate one, and the predators themselves must be under various ecological constraints if they are not to consume all of the hatchlings. Thus, various relatively subtle land-use patterns in the California desert areas (especially the Mojave) have caused unnatural proliferation of ravens, a native species that often preys upon hatchling and juvenile tortoises. Today, ravens consume so many young desert tortoises in some impacted areas that the capacity of the tortoise population to survive is in question.

Similarly in the Galápagos, introduced black rats are present on many of the islands where the tortoises live, but in most cases the rats themselves have predators, including feral cats and dogs, as well as native owls of two species, and Galápagos hawks. On such islands, some natural recruitment of tortoises occurs. On the other hand, on Duncan Island, the only introduced mammal is the black rat, whose populations probably cycle massively under constraints of climate and food availability, but that can reach high populations quickly when conditions are good (Hoeck, 1984). The rats' only predator is the relatively scarce Galápagos hawk. Under these conditions, it appears that natural recruitment of the Duncan tortoise can almost never occur, and the population, despite remarkable longevity of the adults, would have been doomed without the intensive egg-rescue and head-starting program undertaken by the Galápagos National Park and the Charles Darwin Research Station. This effort has spelled new hope for a population of tortoises that seemed to have no future; perhaps it also signals that there is even hope for mankind.

HISTORICAL OVERVIEW

The voluminous literature on Galápagos tortoises, recently inventoried by Beaman (1985), can be divided into four chronologically defined phases. These phases are: Discovery (1535–1840), Taxonomy (1824–1917), Rescue (1925–1955), and *In Situ* (1957–present). The phases reflect the historical development of human attention to the giant tortoises of the Galápagos Islands. Tortoises are known or suspected to have occurred on 11 of the 13 major islands in the archipelago (see Map 1 on next page).

Discovery Phase

The first phase, and the longest, may be termed the "Discovery" phase (1535 to 1840). From the discovery of the archipelago by Bishop Tomás de Berlanga in 1535 to the epochal visit of Charles Darwin exactly three centuries later, the presence of gigantic tortoises in the islands was documented by many, including Dampier (1697), Colnett (1798), Mitchill (1815), Delano (1817), Fitz Roy (1839), and Darwin (1839). Berlanga's own observations on the islands, including the tortoises, were transmitted in a letter to the Holy Roman Emperor, Carlos V, in a letter dated 26 April 1535, and published in English translation by Slevin (1959).

Many of these authors during the "Discovery" phase gave insight into the natural history, abundance, distribution, and economic or subsistence value of the giant tortoises, and gave at least anecdotal information on their inter-island differences. However, it was the general practice at the time to refer to all island giant tortoises, worldwide, as a single species, called *Tortue des Indes* by the pre-Linnaean authority Perrault (1676), and *Testudo indica* by Schneider (1783).

Taxonomic Phase

Overlapping only slightly with the "Discovery" phase was a second period (1824 to 1917) that may be termed the "Taxonomic" phase. This period started with the first proposals of scientific names for Galápagos tortoises — *Testudo californiana* and *T. nigra*, by Quoy and Gaimard (1824a, 1824b), and *T. elephantopus* by Harlan (1827). During this phase, European and American zoologists, usually associated with major museums, described and gave names to the specimens that arrived in their institutions. These were often from unknown or vague sources; the provenance of none of the type specimens of the above-named taxa can be traced to a particular island, nor can it be deduced from the morphology of the specimens, since all were juveniles. Indeed, the juvenile that served as holotype for both *Testudo californiana* and *T. nigra* (MNHNP 9550, carapace length 26.8 cm, Figs. 6-8) was reported to have come from California, and to have been donated, as a living animal, to M. de Freycinet by M. Meek, captain of the American ship *Boston Eagle* (*l'Aigle de Boston*). The donation was made in Hawaii (*îles Sandwich*). The alleged California origin of the specimen may have been supposition based upon the superficial similiarity of a young Galápagos tortoise to an adult of the California desert tortoise, *Gopherus (Xerobates) agassizi*, although it must be admitted that there is no evidence that Quoy and Gaimard were familiar with this species.

Nevertheless, where the type material still exists, questions as to the precise origin of these specimens may not remain forever unanswered as biochemical and genetic techniques and DNA and protein analyses become ever more sophisticated.

The numerous writings of Günther (1875, 1877, 1896, 1898, 1902) and Rothschild (1896, 1901, 1902a, 1902b, 1902c, 1902d, 1903, 1915a, 1915b, 1925, 1928) contributed much during this phase, although neither could resist the urge to name new species based on inadequate material — often single specimens of unknown origin. Günther's 1877 monograph on the giant tortoises, beautifully illustrated and offering detailed descriptions of the osteology of many specimens, stills falls short of providing real insight into the distribution and relationships of the different Galápagos forms.

This task fell to John Van Denburgh (1914) who published by far the most comprehensive account of the Galápagos tortoises ever written. Copiously illustrated with photographs of specimens from standardized angles and based upon newly-collected material totalling 256 specimens of precisely known origin (collected on the *Academy* expedition of 1905–06, largely by Rollo Beck and Joseph Slevin), it stands as a work of enormous value even today. Moreover, since the current prohibition on collecting live specimens of Galápagos tortoises is likely to continue, and since some of the island populations are extremely rare or extinct, there is no possibility of duplicating Van Denburgh's study series, housed today in the California Academy of Sciences in San Francisco.

Yet both Van Denburgh and Günther — and also the talented Georg Baur — fell into the trap of attempting to determine the provenance of certain specimens of unknown origin by a process of logical argument. Specimens in hand (often single skulls or shells) were compared with isolated specimens of known origin, accounts by ancient mariners

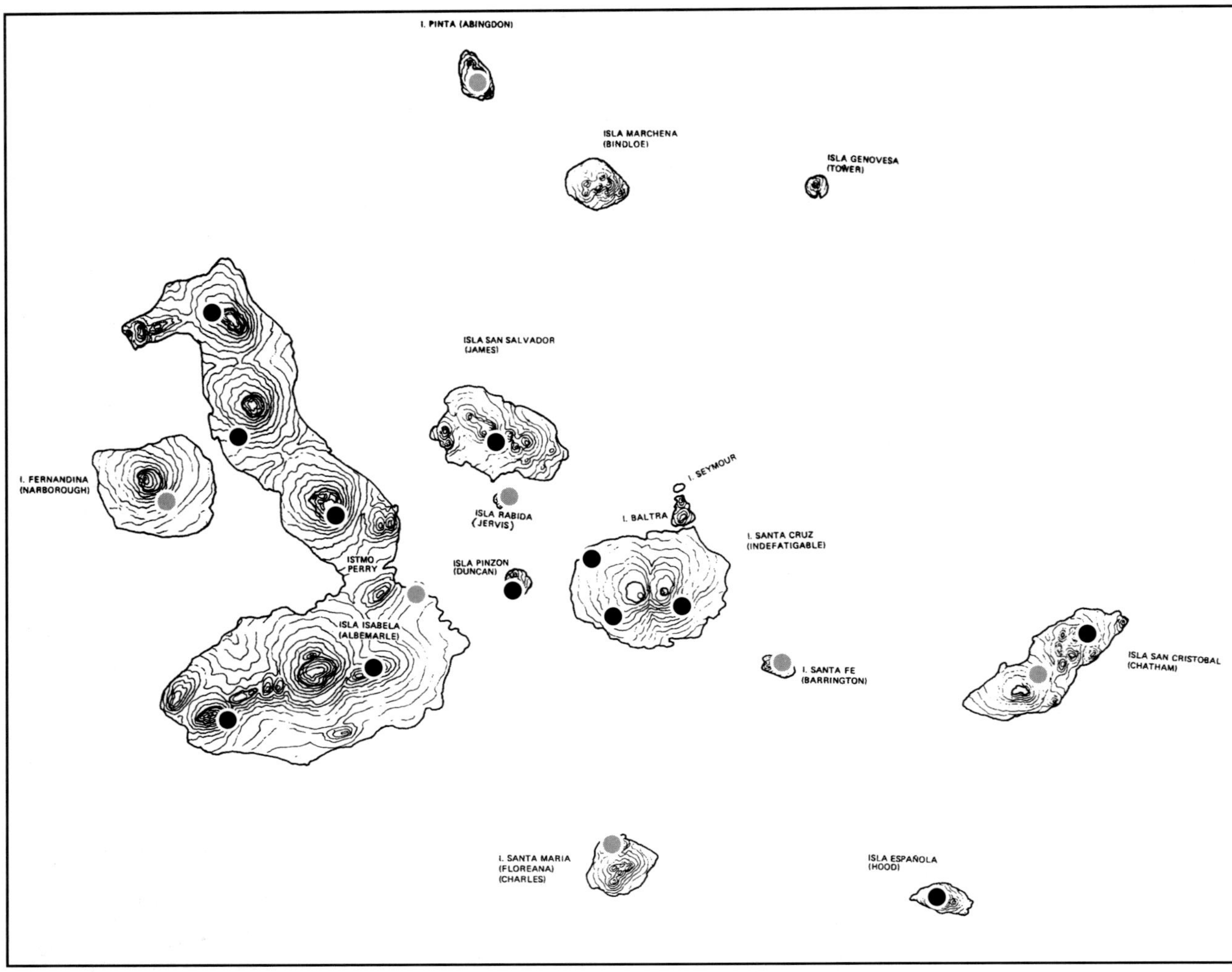

Map 1. Galápagos Islands. Tortoises are known or suspected to have occurred on 11 separate islands: Narborough (Fernandina) (probably extinct), Albemarle (Isabela) (five separate populations, distributed on the separate volcanoes), Abingdon (Pinta) (probably extinct in the wild, one individual male alive in captivity), James (San Salvador), Jervis (Rábida) (extinct, possibly introduced), Duncan (Pinzón), Indefatigable (Santa Cruz), Barrington (Santa Fé) (extinct, possibly introduced), Charles (Santa María or Floreana) (extinct), Chatham (San Cristóbal) (two separate populations, one extinct), and Hood (Española). Black dots represent extant populations, lightly shaded dots represent extinct or extirpated populations.

were consulted, and rumors as to when certain island forms had become extinct were discussed — the intellectual pot was stirred and fermented and, with an almost visible flourish, *magister dixit*, and the origin of the specimen was declared. Such guesses were presented as gospel, or at least as sufficient basis for making nomenclatural decisions often involving the naming of new taxa, even though it is clear from Van Denburgh's numerous photographs — and also from the inability of modern zoo curators to identify the tortoises in their charge — that the variation is enough to make such guesswork exceedingly unwise. Yet when he had adequate material, Van Denburgh was capable of great and cautionary wisdom, as in his comments about Galápagos tortoise skulls:

"In a series of 24 skulls from Vilamil, Albemarle, I find all of the variation which Günther mentions; and upon careful comparison of this series with one skull from Hood Island (No. 8125), one from Indefatigable (No. 8381), one from James (No. 8105), three from Duncan (Nos. 8378, 8379, and 8380), four from Chatham (Nos. 8127, 8128, 8130, and 8131), and one from Iguana Cove (No. 8179), I can find no constant differences in the skulls of the various races. In the skulls from Vilamil, the frontal region may be flat or somewhat convex. The occipital spine may be short or long, not reaching the posterior borders of the mastoid processes or projecting far behind them, and may or may not rise much above the level of the skull. There is much variation in the shape of the tympanic case and cavity. The fossa in front of the occipital condyle may be deep or very shallow. The tuberosity for the temporal muscle may be quite small or very largely developed. The nasal opening may be as high as broad, or broader than high. The palatal region varies much in shape. It may be narrow or broad, and the pterygoid edges may be sharp or blunt. The alveolar ridges also vary in position and degree of development. We may safely say that no constant differences exist among the skulls of the various races of Galapagoan tortoises."

Although Rothschild published short papers in 1925 and 1928, Samuel Garman's 1917 revision of the Galápagos tortoises, based upon the collections of L. Agassiz and G. Baur at the Museum of Comparative Zoology at Harvard

University, essentially represents the final flourish of the "Taxonomic" phase of Galápagos tortoise documentation. To this day, this idiosyncratic work must be considered the last comprehensive attempt to establish a nomenclature for Galápagos tortoises. That such an uneven effort should have withstood formal challenge for three quarters of a century is little short of incredible. Yet, in the field of giant tortoise biology, the cycles of assertation, argument, and counter-argument may be as slow as the tortoises themselves. Witness, for example, the field account of the Indian Ocean Rodrigues tortoises (*Cylindraspis*) offered by Leguat (1708), following his sojourn on Rodrigues during the years 1691–94. It took over two centuries for Atkinson (1922) to question Leguat's work by the devastating technique of arguing that Leguat had never existed; the book, Atkinson claimed, was a fictitious account by an armchair travelogue writer. More than half a century more was to pass before North-Coombes (1979) defended both Leguat's existence and his observations in a meticulously researched and extraordinarily thorough thesis.

Rescue Phase

Beginning approximately with Beebe's celebrated 1925 book *Galápagos — World's End*, and continuing with the writings of Charles Townsend (1924a, 1924b, 1925a, 1925b, 1928a, 1928b, 1931a, 1931b, 1932, 1942), Ralph De Sola (1929, 1930), and others, a third or "Rescue" phase of Galápagos tortoise writings ensued. With the assumption that the tortoises were so rare and so stressed in their native habitat (by direct take by settlers, and by competition or predation from feral mammals) that their survival was unlikely, it was considered imperative to transfer them to zoos, where they would be encouraged to reproduce. The efforts started shakily, with Beebe's collection of a single Duncan Island tortoise that died after a week, but Townsend in 1928 was able to collect a large series of 180 mostly juvenile specimens, largely from southern Albemarle Island. Some of these later reproduced in zoos, and a few survive to this day (Fig. 9).

The captive groups were also the subject of behavioral studies by Evans (1949), Evans and Quaranta (1949, 1951), and others. During this phase, only De Sola (1930) gave a new name (*Testudo vandenburghi*) to the live tortoises he collected, and since this very brief paper did not identify type specimens nor features by which the new species could be distinguished from others, confusion rather than clarification was the result.

Typical of the writings of this period was the observation of Banning (1933), who, in justifying the taking of a

Figure 6. Original plate of the holotype of *Testudo nigra* (MNHNP 9550) from Quoy and Gaimard, 1824b. Photo courtesy of Bibliothèque du Muséum de Paris.

Figures 7 and 8. Ventral and lateral view of holotype of *Testudo nigra* (MNHNP 9550) as it exists today. Photos courtesy Jean Lescure, Muséum National d'Histoire Naturelle.

series of 18 Galápagos tortoises (12 actually obtained from Guayaquil on the mainland rather than from the islands themselves) to San Diego by the Hancock expedition of 1933, commented: "If the 'galapago' — the prodigious giant tortoise from which the islands took their name — is to survive, it is scarcely probable that his native soil will provide the environment."

The "Rescue" phase finally achieved fulfillment with the announcement of successul captive breeding of Galápagos tortoises in the zoos of San Diego (Staedeli, 1958; Shaw, 1959, 1961a, 1961b) and Honolulu (Throp, 1969, 1975). The realization followed, however, that captive breeding would best be undertaken — where necessary — in the islands themselves, utilizing breeding stock or known and homogeneous island material. The "Rescue" phase reached full cycle when two representatives of the rarest island forms, a Duncan tortoise in the Bronx Zoo (Fig. 10) and a Hood Island tortoise in the San Diego Zoo (Fig. 11), were returned to the islands (Bacon, 1978; Pritchard, 1979). Significantly, in both cases these individuals were of wild-caught stock rather than captive-bred. Nevertheless, some individuals of the original 1928 collection still survive in US institutions, and in recent years there has been excellent captive reproduction, e.g., at the Gladys Porter Zoo in Brownsville, Texas, and at the Life Fellowship Facility in Seffner, Florida (where 85 hatchlings were produced in 1988–89, with some females nesting five times in a season). As of 1994, Life Fellowship had hatched and kept alive a total of 411 Galápagos tortoises.

In Situ Phase

During World War II and for a decade thereafter, opportunities to visit the Galápagos were few (except for those on Baltra Island on US military service), and no field studies were undertaken. However, starting around 1957 with the UNESCO/Life Magazine Galápagos Expedition under the scientific leadership of I. Eibl-Eibesfeldt and Robert Bowman, a new phase was initiated, continuing to the present, that may be called the "*In Situ*" phase. Events now unfolded rapidly, leading not only to the construction of the Charles Darwin Research Station and the establishment of the Galápagos National Park of the Republic of Ecuador, but also to a vast and increasing volume of literature based upon field studies of the Galápagos tortoises. Some conspicuous contributions during this phase were those of Eibl-Eibesfelt (1959), Snow (1964), MacFarland et al. (1974a, 1974b), MacFarland and Reeder (1974), Schafer (1982, 1983), Fowler (1983), and Cayot (1987), although these titles represent only a small fraction of those generated from 1958 to date.

Details of the "*In Situ*" phase conservation activities are given in the Survival Status section later in this work. The essence of the strategy devised by the experts at the Charles

Figure 9. Scene in Houston Zoo, Texas, 1970. Zoogoers admire a large adult male tortoise from Albemarle Island (Isabela), "rescued" as a juvenile from the Iguana Cove (Cerro Azul) area in 1928 by Charles Townsend.

Figure 10. Adult female Duncan Island (Pinzón) tortoise, captured by Charles Townsend for the Bronx Zoo in 1928, and returned to Duncan in 1972. The white head is typical both of old Galápagos tortoises and of representatives of the saddlebacked races.

Figure 11. Adult male Hood Island (Española) tortoise, resident in the San Diego Zoo for over 40 years until its return to the Galápagos in 1977, where it is now a contributor to the successful captive breeding program for this endangered population. May 1970.

Figure 12. Juvenile captive-hatched Duncan Island tortoises being head-started at the Charles Darwin Research Station. August 1986.

Darwin Research Station and the Galápagos National Park was a recognition that the tortoises were menaced by a mosaic of stress factors, the precise combination of which was unique to each population. Thus, an inventory of different techniques would be required, with the least manipulative (and the least expensive) being applied where such techniques would be adequate, or where population sizes were sufficient to absorb some losses, but with progressively more ambitious techniques applied where necessary.

The stress factors identified included capture of tortoises by smugglers for sale overseas, slaughter of tortoises by settlers, food competition by feral herbivores (especially goats, but also cattle and burros), predation on primarily subadult tortoises by dogs, predation on juveniles and eggs by pigs, and predation on hatchlings by introduced rats, or by native species including the Galápagos hawk. Furthermore, some populations were so depleted that even sporadic, accidental mortality of adults could not be tolerated, or reproduction had effectively ceased because encounters between males and females no longer occurred.

For some populations, the solution was a combination of law enforcement, education, and control or, where possible, elimination of feral mammals. However, in the case of the tortoises of Duncan Island, where feral rats consumed every hatchling, it became necessary to collect eggs from the wild, incubate them artificially, and head-start the offspring for several years before release on their ancestral island (Fig. 12). Transfer of eggs to artifical incubation and head-starting of the resulting young was also practiced for the tortoises of James and Chatham islands, where the eggs and young were threatened by feral pigs and dogs, respectively. Only in the case of the extremely depleted tortoises of Hood Island was the most labor-intensive technique of all — the capture and conveyance of the entire wild population to a captive-breeding facility — found to be necessary.

The ultimate challenge was presented by the discovery of a single surviving male tortoise on Abingdon Island in 1972. The animal has resided in captivity at the Charles Darwin Research Station for the ensuing 24 years, but attempts to mate him with females from Volcan Wolf (Albemarle Island) have failed, and if any of his genes are to survive, technological remedies will have to be sought and applied.

NOMENCLATURAL STATUS

In the sections that follow various aspects of the nomenclature of Galápagos tortoises are discussed. Many problems exist, from the actual scientific names used to the validity and distinctness of the various populations or forms. The currently commonly used species name for all Galápagos tortoises, *Geochelone elephantopus*, whose usage (with separate subspecies for the various island forms) dates only back to Mertens and Wermuth (1955), is not the oldest available name for the species and therefore needs to be corrected. Many populations bear inaccurate or erroneously allocated names, several named populations need to be synonymized, and some distinct forms need to be named. The first section below reviews and critiques the last major taxonomic work regarding Galápagos tortoises (Garman, 1917). The current status of the nomenclature is then analyzed in the second section, followed by a summary of recommendations.

Garman's *The Galapagos Tortoises*

Samuel Garman [1843–1927] was 73 years old when he published *The Galapagos Tortoises* in 1917 — his only publication on tortoises. A famous and distinguished senior herpetologist and heir to the Agassiz mantle at the Museum of Comparative Zoology (MCZ) at Harvard University, Garman had made notable contributions to the study of sea turtles, having described two new species (*Lepidochelys kempi* and *Chelonia depressa*) in 1880. These remain the only valid species of sea turtle to have been described since Bocourt's description of *Chelonia agassizi* in 1868. Yet Garman was a highly eccentric and less than meticulous worker. His odd and secretive working habits are described by Adler (1989). Garman's type series of *Chelonia depressa* included a perfectly typical juvenile green turtle, *Chelonia mydas* (Barbour, 1914), while in his description of *Lepidochelys kempi* he repeated purely hearsay information to the effect that this species nested in the Florida Keys from December to January (in fact, it is virtually confined as a nesting species to Tamaulipas, Mexico, and it only nests in the spring months). Garman also, in describing the new and valid mud turtle species *Kinosternon baurii* in 1891, reported this species not only from Florida, where it indeed occurs, although no longer at the type locality of Key West (Lazell, 1989), but also from Cuba (where no mud turtles occur).

Garman's 1917 work on the Galápagos tortoises, published about a quarter century after the close of the author's primary herpetological output, is also somewhat puzzling, inconsistent, and superficial. Although chronologically the last major paper in the "Taxonomy" phase of Galápagos tortoise documentation, it is certainly not the last word. Perhaps it was more common in 1917 than today for the contributions of distinguished elder statesmen of science to receive only superficial critical comment by presumably younger and certainly deferential technical reviewers, especially when they were submitted to journals over which the authors had administrative authority. Certainly this seems to have been what happened with Garman's work.

Garman's stated stimulus for writing his review was the large collection of Galápagos tortoise material — mainly shells — that had accumulated at the Museum of Comparative Zoology. But it also hearkened back to a much earlier phase of Garman's life — to 1872, when he was in Wyoming collecting fossils for Edward Drinker Cope. Cope fired Garman for making unreasonable salary demands, and Garman drifted westward, arriving in San Francisco in time to see Louis Agassiz and his party arrive from the Galápagos on the ship *Hassler*. Agassiz recognized Garman's potential on the spot, and offered him a place at Harvard College as his pupil. Thus, the tortoise specimens and Garman himself arrived at Harvard at around the same time.

Perhaps Garman's work was also motivated by a little transatlantic rivalry — Rothschild (1915a, 1915b) had just published a catalogue and account of the large collection of giant tortoises in his personal museum at Tring. Yet Garman's material in many cases had dubious or absent collecting data, and even when the island of origin was recorded — as with the juveniles from Charles Island (= Santa María or Floreana) obtained by Agassiz on the *Hassler* expedition in 1871–72 — the tortoises were purchased from locals who had presumably brought them to this island from elsewhere, the local tortoises having become extinct very shortly after Darwin's visit in 1835 (Seemann, 1846, in Van Denburgh, 1914). How or why Garman concluded (p. 266) that the *Hassler* specimens, too small to show any distinctive island characteristics, were referable to four different species, is completely unclear.

Had Garman's paper appeared before Van Denburgh's 1914 monograph, the purpose — of doing the best possible with available material — would have been evident. As a subsequent paper, however, it seems to take scientific understanding of the Galápagos tortoises a step backward. Garman's material might have justified the publication of a catalogue, but the material in the MCZ collection did not warrant using it as virtually the sole basis for a review of the group. Van Denburgh (1914) had recognized fourteen species of Galápagos tortoise, several of them (*Testudo hoodensis*, *T. chathamensis*, *T. darwini*, and *T. phantastica*) described by himself a few years earlier (Van Denburgh, 1907). Garman assigned *T. darwini*, *T. wallacei* (which he incorrectly attributed to Van Denburgh rather than to Rothschild), and the unnamed form that Van Denburgh reported from Cowley Mountain (= Volcan Alcedo) to the synonymy of *Testudo nigrita* Duméril and Bibron, 1835. He synonymized *T. hoodensis*, *T. becki*, and *T. phantastica*, as well as the earlier species *T. ephippium* and *T. abingdoni*, with *Testudo elephantopus* Harlan, 1827. The latter name was used for a mixed assemblage of saddlebacked forms, even though its holotype was a juvenile of unknown island origin, and irretrievably lost. (A.E. Brown, *in* Van Denburgh, 1914, had located the cleaned leg bones of one side of a tortoise, and a part of the legs of the other side still bearing some dried skin, in the collection of the Philadelphia Academy of Sciences; these were probably all that remained of the type of *T. elephantopus*).

Furthermore, Garman (p. 269) did not help those seeking his actual views by listing *T. abingdoni*, *T. becki*, and *T. duncanensis*, all expressed in binomial form, as "varieties" of *T. elephantopus*, while later (p. 294) he lists *T. ephippium*, *T. abingdoni*, *T. becki*, *T. hoodensis*, and *T. phantasticus* [sic] as "so-called species." The name *T. duncanensis* was quoted only once, *in passim*, and is a *nomen nudum*. We can suspect from Garman's quotation of Baur's (1889) insightful argument (pp. 295–296) that the holotype of *T. ephippium* was not in fact from Duncan Island but was instead an Abingdon tortoise (generally known since 1877 as *T. abingdoni*), and that he intended *T. duncanensis* as a new name for the Duncan Island tortoise now that *T. ephippium* was unavailable; but this is assumption only.

Figure 13. Dorsal view of cotype of *Testudo rotunda*, Latreille, 1802, a hatchling Rodrigues tortoise. Photo by R. Bour.

While disposing of Van Denburgh's new names in this summary but ambiguous fashion, Garman presented several new names of his own. Although aware (in his very first paragraph) that Galápagos tortoise populations showed such individual variation as to require analysis of substantial series to characterize them adequately, he based his new species *Testudo clivosa* on a single, unremarkable bony shell with a few scutes still attached. The island of origin was unknown. Similarly, no locality was given for either the holotype or the only other reported specimen of the new species *Testudo typica*. The two specimens (Plates 22 and 34) seem to differ considerably in relative width and roundness, while the similarity of the lateral profiles leads one to suspect error in the measurements (p. 286) that indicate that the two shells differ by five inches in straight-line carapace length but by only one inch in curved length. The length of the larger (MCZ 5260) is incidentally given as 29 inches on p. 286 but as 28.5 inches in the caption to Plate 34 — a careless inconsistency, but an understandable error nonetheless; repeated measurements of large turtle shells rarely yield identical results.

Garman also proposed the name *Testudo macrophyes* as a substitute name for the Tagus Cove tortoise, described earlier as *T. microphyes* by Günther (1875, 1877) on the basis of several specimens of known origin collected by J. Cookson, but with a holotype (BMNH 75.12.29.1) of unknown origin that Garman concluded was abnormal, aged, and dwarfed. Perhaps Garman may have been justified in dismissing this tortoise as a suitable holotype for the Tagus Cove tortoise, but in view of his condemnation of the normality of the specimen it is surprising that he retained the name *T. microphyes* as a valid species, apparently from southern Albemarle, of which *T. güntheri*, *T. galapagoensis*, and even *T. chathamensis* — from the far side of the archipelago — were listed as synonyms. What was not particularly surprising was that Garman's new names (i.e., *T. macrophyes*, *T. clivosa*, *T. typica*, and *T. duncanensis*) were never used again by anyone.

A few comments on points raised by Garman in the introductory section of his work (pp. 261–270) are appropriate. Particularly enigmatic is the argument presented at the top of p. 262, in which Garman argues that the Galápagos organisms reached the islands either by "origin in place," or by "accidental importations;" his discounting of the latter possibility suggests that he subscribed to the former theory.

Figure 14. Adult male James Island (San Salvador) tortoise, in the western highlands of James. Note remnant *Scalesia pedunculata* var. *pedunculata* trees in a habitat largely converted to grassland or *pampas*. February 1971.

Had the phrase been "evolution in place" rather than "origin in place," the argument might have been more acceptable; as it stands it appears to argue for the independent creation of finished organisms in the Galápagos Islands — a view that reflected, perhaps, the influence of Louis Agassiz (or, through Agassiz, even of Georges Cuvier).

On pp. 262–263, Garman reflected upon the Spanish and English names by which the different Galápagos islands were known, suggesting that the Spanish names were older, that Ecuador still clung to them, and that they would ultimately prevail. Yet in fact the English names had been established by seventeenth century English pirates such as Dampier and Cowley, whereas the Spanish names, almost all of which reflect some aspect of Columbus' voyage to the Americas (e.g., the ships *Pinta* and *Santa María*; Queen *Isabela* of Spain; *San Salvador*, the island on which Columbus first landed; *Genovesa*, Columbus' birthplace, etc.) were given on the occasion of the fourth centenary of Columbus' first voyage to the Americas (i.e., in 1892) only 25 years before Garman's work was published. However, some Spanish names are older, James Island having been renamed *Olmeda* and Charles Island *Floreana* by Col. Ignacio Hernández in 1832. In this work, I elect to use the English names (but also give Spanish equivalents in the Summary and Survival Status sections), not just for reasons of chronological priority (an important consideration to taxonomists), but also because this paper is not intended as a field guide, but rather as an academic contribution and commentary upon the earlier biological literature on the Galápagos tortoises, which in general (exclusively, in earlier works) utilizes these names. For exhaustive synonymies of Galápagos Island names, see Black (1973) and Woram (1989).

Mysterious indeed is Garman's theory that the Galápagos tortoises had been introduced by man. A century earlier, it had been thought that all island giant tortoises, in both the Indian and Pacific Oceans, were referable to a single species, *Testudo indica*, which would naturally lead one to seek evidence of human introduction to the far-flung islands on which the species was supposed to occur. Darwin himself subscribed to this view in the first edition of his *Journal of Researches* (1839), as did Fitz Roy (1839), but he corrected himself in a footnote to his first edition, and in the text of his second edition (Darwin, 1845), in which he referred to the Galápagos tortoises as "*Testudo nigra*, formerly called *Indica*," but also commented that there were "two or three species or races in the Galápagos Islands." The key influence in this revision of opinion had been Gabriel Bibron, who had addressed the Zoological Society of London on 28 February 1838, and who had insisted that he had seen at least two species of tortoises from the Galápagos Islands alone (Sulloway, 1982). Nevertheless, as late as 1870, John Edward Gray wrote "The black tortoises of Asia that have been called *Testudo indica*, which are found spread over all parts of the Asiatic region, also on the islands of the east coast of Africa, and in California and the Galápagos... Modern writers on the subject have united these into a single species under the name *Testudo indica*" (Gray, 1870).

Agassiz (1857) failed to note Darwin's 1845 correction, referring to the scientific name of the "Gallapago turtle" as *Cylindrapis* [sic] *indica*, i.e., still utilizing the "universal" specific epithet for insular giant tortoises. *Cylindraspis* had been proposed as a new genus by Fitzinger (1835), with subsequent designation of the Rodrigues saddleback tortoise (*Testudo vosmaeri*) as the type species (Fitzinger, 1843).

Garman claimed that several distinct species were peculiar to various of the Galápagos Islands, but this leaves entirely unanswered the question of the original habitat of what he considered may have been "introduced" populations.

Actually, the superficial similarities between the Indian Ocean and the Galápagos tortoises — large size, unpatterned dark shells, enlarged shell openings, long necks, absence of osteoderms in the limb scales, shortened plastra — are striking, and would surely lead a modern cladistic phylogeneticist unschooled in zoogeographic considerations or other special knowledge of the group to consider them a true (monophyletic) clade, even if not a single species. Even the variation within the groups — for example, the development of saddlebacked shells among some of the Galápagos forms and also in the Indian Ocean species *Cylindraspis vosmaeri* — is remarkably parallel; saddlebacked shells appear to be a shared derived character or synapomorphy if ever there was one.

As a remarkable latter-day commentary upon the parallelism between the Galápagos and the Mascarene tortoises, Roger Bour advises (*in litt.*, 19 December 1989) that the two hatchling tortoises, preserved in spirit in the MNHNP collections (Fig. 13), on which Latreille (1801) based his description of *Testudo rotunda*, are in fact Mascarene tortoises, probably from Rodrigues, whose tortoises were still marginally extant when Latreille wrote his account. In fact, the same specimens had earlier been figured by Lacépède (1788) — proof positive that they were collected in or before 1788. They were not Galápagos tortoises, as Bour himself (1980) had earlier identified them, and as they had also been considered by those few twentieth century systematists who had taken any note of the existence of this name. Bour's comparisons had been hindered by the fact that no collection has unequivocal Mascarene tortoise hatchlings, and the Paris collection lacked hatchlings from the Galápagos also. Moreover, the name *T. rotunda* offers little clue (rotundity is a characteristic of all tortoise hatchlings); and the actions of the reviewers in the years following 1801 were decidedly confusing.

The name is an important one — if it were based upon a Galápagos tortoise, it is the oldest available name for a

Figure 15. Adult male James Island tortoise, from the western highlands of James. The anterior elevation of the carapace is greater than is typical of this population. February 1971.

member of that group, has extant type material, and is not a *nomen oblitum*. Schweigger (1812) noted that Lacépède had considered this species to be a synonym of the European pond turtle, *Testudo orbicularis* or *Emys europaea* (both of which names he rendered in the genitive rather than the nominative case), but he disagreed vigorously with this conclusion, observing that the species was clearly entirely terrestrial, and he utilized the name *Testudo rotunda* in unaltered form, as a valid species. Nevertheless, Merrem (1820) continued to consider *Testudo rotunda* to be a synonym of *Emys orbicularis*, an error that has survived to recent times (Wermuth and Mertens, 1977).

Schweigger's description of the holotype of *T. rotunda* included several useful diagnostic characters. For example, the statement "*bracteae marginales viginti tres, anticae aequales*" (marginals 23, the front ones equal) indicate that the nuchal was absent, as in both Galápagos and Mascarene but not Seychelles tortoises. On the other hand, the statement "*sternum...in duodecim areas divisum*" (12 plastral scutes) would favor the Galápagos rather than the Mascarene tortoises; but this is not a necessary conclusion. Although most adult Mascarene tortoises have a single gular (or fused gulars), giving eleven plastral scutes, certain subfossil plastra from Mauritius attributed to *Cylindraspis triserrata* have the gulars paired and indeed anteriorly extended and divergent (Bour, 1985). The gular is also occasionally divided in the Réunion *Cylindraspis* (R. Bour, *pers. comm.*), and is dorsally divided in the young of *Astrochelys yniphora* and in the holotype of *Cylindraspis graii*. Furthermore, turtle species are known in which a median epidermal shell element is divided in the young, but by a process of gradual fusion becomes ontogenetically undivided. This occurs especially in areas where the scutes grow by distal rather than proximal accretion of new substance, as is the case with the anal scutes of certain *Cuora* species, or the supracaudal scute of *Peltocephalus*.

Some of the known adult shells of the Rodrigues tortoise show actual separation of the anteriormost pair of marginal scutes, with the first vertebral scute forming part of the anterior margin of the carapace. However, this is probably again an ontogenetic character, associated with the development of the raised, saddled anterior part of the shell, and was apparently exacerbated also in adults by damage to this area possibly resulting from intraspecific biting or fighting, with resultant loss of shell material and formation of scar tissue (pers. obs.). Important in Schweigger's description is the reference to the spur on the end of the tail ("*Cauda crassa exserta, apice corneo*"), a massive tail spur being a feature of adult male Mascarene tortoises.

But despite this confusion, the skulls alone are sufficiently different to justify separation of the three giant tortoise stocks at the generic level — *Geochelone* for the Galápagos forms,

Figure 16. Old adult male tortoise under a barbed wire fence in cattle pasture near Santa Rosa, Indefatigable Island (Santa Cruz). The peeling texture of the carapace scutes is probably a result of infestation with fungal and other microorganisms in this very humid environment. July 1986.

Figure 17. Holotype (CAS 8127) of *Testudo chathamensis*, Van Denburgh, 1907, from southwestern Chatham Island (San Cristóbal). Photo from Van Denburgh (1914), plate 64, figure 2.

Aldabrachelys for the Aldabra-Seychelles-Madagascar tortoises, and *Cylindraspis* for the Mascarene forms.

On pp. 264–266 Garman discussed his belief that part of the problem in classifying Galápagos tortoises lay in the likelihood that tortoises were extensively transported from one island to another, resulting in hybrid, morphologically confusing populations. Such inter-island transport of tortoises undoubtedly occurred from time to time, but on nothing like the scale of tortoise transport between islands of the Indian Ocean. The morphology of the surviving populations does not show evidence of genetic contamination of the different island stocks, although in a few cases (such as that of the Volcan Wolf tortoise) the morphological variation seen today is somewhat difficult to explain. The main element of uncertainty presented by the inter-island transportation possibility relates primarily to the extinct forms — to knowing whether or not dried bones or subfossil remains found on such islands as Charles, Barrington, and Jervis are truly referable to endemic, extinct subspecies rather than to tortoises brought from other islands.

On pp. 267–269, Garman entered a detailed discussion of the topic of sloughing, with examples taken from various unrelated reptiles, including rattlesnakes and the flatback sea turtle. Some freshwater turtles, especially emydids and chelids, may slough the outer layer of a growing scute in a single piece, but this does not occur with Galápagos tortoises. The question of why the scutes of Galápagos tortoises differ in texture and surface sculpturing is a complex one, discussed to some degree by Fritts (1984), but Garman's conclusions that the carapacial texture of Galápagos tortoises is a product of the sloughing of entire scutes does not appear to be based on actual observation, nor is such sloughing known to occur in these tortoises. The correct interpretation is probably that smooth-shelled individuals became mechanically abraded to a smooth condition after they reached maturity, although perhaps smaller, chronically undernourished tortoises may also show smooth shells at times. The former hypothesis may explain in part why adult female Galápagos (and Aldabra) tortoises are usually smooth-shelled. However, it is unlikely to be the complete explanation, since abrasion by contact with the plastron and the forelimbs of the male during copulation would not affect all parts of the carapace equally. The question is not peculiar to Galápagos tortoises; many long-lived chelonian species, especially terrestrial ones (including such genera as *Gopherus* and *Terrapene*), may include both smooth-shelled adults and others with well-sculptured annuli.

Figure 18. Young adult male tortoise surveys the flooding of the lowlands of northeastern Chatham Island (San Cristobal) during the El Niño episode of 1972.

Figure 19. Adult female Volcan Wolf tortoise nesting in a small patch of soil surrounded by pahoehoe lava inland from Bank's Bay, Albemarle Island (Isabela). February 1971.

Garman's discussion on p. 269 of possible differences between Galápagos tortoises collected in early and in recent years is exceedingly interesting. Fritts (1983) wrote that "preliminary comparison of museum specimens collected during the period 1890–1910 with specimens found living in Galápagos in 1976 and 1977 indicate that population differences are stable and do not vary with short-term climatic or ecological fluctuations." Yet both Garman's and some more modern observations suggest otherwise, and factors more complex than the progression of the Duncan Island tortoise population from a vigorous one at the turn of the century to a senile one (with repatriated young animals present, but none in their middle years) may have been at work. For example, Garman quotes Porter's (1815) account of the round, plump, black, and thin-shelled tortoises of James Island (Fig. 14), whereas a century later Van Denburgh reported the tortoises on this island to be large, heavy, thick-shelled, and somewhat saddlebacked rather than rounded.

Furthermore, my own observations in recent years suggest that some changes have taken place since Van Denburgh's specimens were collected at the beginning of this century. In some cases the changes may relate to alterations in the population structure. For example, a century ago, the tortoises of James and Indefatigable Islands were regularly preyed upon by settlers, and really large, old animals may have been very scarce. Today, the threat is primarily upon the hatchling and juvenile stages (e.g., by rat, pig, or dog predation), and large animals are essentially safe. The adult specimens found today on James (Fig. 15) and on Indefatigable (Fig. 16) include individuals larger than any reported by Van Denburgh, and many may be seen with smooth, unstriated shells, whereas even the adults collected by Van Denburgh showed the shell sculpturing typical of younger adults.

The tortoises reintroduced to Duncan and Hood are growing larger than their parents (Rhodin et al., 1983; Pritchard, 1985). The tortoises of Chatham Island, reported to be flat-shelled by Van Denburgh (Fig. 17), are now significantly saddlebacked (Fig. 18) (indeed, as described below, they are of a different population altogether, although on the same island, and curiously sharing the frequent anomaly of pectoral scutes that fail to meet in the midline, a detail recorded by Van Denburgh, and which I found in 9 of 18 specimens examined in the field). And the tortoises of Volcan Wolf (Fig. 19), described as saddlebacked by Van Denburgh (although with considerable variation in the degree of anterior elevation of the carapace) are now known to include both domed and flattened individuals as well as a wide variety of saddlebacked specimens.

The causes of these differences remain conjectural. Some hypotheses are presented elsewhere in this paper. The very large size of the adult male tortoises on James and Indefatigable today may stem from the change in population structure discussed above, but also may be associated with the progressive conversion of large areas of the highlands of these islands into lush savannahs or pasturelands locally known as *pampas* (Fig. 14). These have developed as a result of human land-clearing for pasture on Indefatigable Island, and of feral goats, present in tens of thousands, destroying large amounts of native upland vegetation on James. The ecological changes wrought have been serious and many plant species may disappear. But, as a curious secondary result, the grasslands, that have developed at cool, high altitudes at which tortoises are rarely in danger from excessive insolation even when no shade is available, provide excellent tortoise pasture, and the individuals that graze there are growing large and fat accordingly. There seems to be no evidence for the speculation of Fritts (1984) that cool, moist (i.e., upland) habitats may result in a slowing of growth of all adults (male and female), although there may be evidence that some individuals do not move to such altitudes until already well-grown.

Peter and Rosemary Grant and their students and co-workers studied the finches (Geospizinae) of the island of Daphne Major for more than two decades (Weiner, 1994). Their extraordinarily detailed observations indicated that, among these small birds, reproduction could occur at an age of less than three months in especially rainy years, and that the populations on Daphne showed fluctuations or progressions, especially in beak dimensions and overall body size, on a year-to-year basis. Detectable microevolution at such a tempo would not be expected among Galápagos tortoises, that take at least two decades to reach first maturity and that show great individual resistance to stressful conditions such as prolonged drought (conditions that quickly decimate finch populations, leaving presumably only the best-adapted individuals as survivors). Nevertheless, the same phenomenon could conceivably be occurring in the various populations of Galápagos tortoises on a very protracted scale — measurable only over decades rather than over months or years.

Garman went to some length to redescribe the species *Testudo tabulata*, basing his description largely upon an allegedly Puerto Rican specimen of which he provided photographs. Although current opinion suggests that this modern species, while related subgenerically to the Galápagos tortoises, is certainly not ancestral to them, Garman (p. 272) thought otherwise. The name *Testudo tabulata* was proposed by Walbaum (1782), but this work was not consistently binomial in its nomenclature, and the Latinized names included are only cited *in passim*, as translations of vernacular names (Vanzolini, 1977), and are thus not considered available for nomenclatural purposes. The species long known as *Testudo tabulata* is in fact a composite of two species, *Geochelone denticulata* and *G. carbonaria*, whose distinctness was overlooked by most twentieth century authorities until the matter was finally clarified by Williams (1960).

Moreover, the specimen selected by Garman for detailed description was a confusing one. Its stated origin of "Puerto Rico" is clearly mistaken. While the specimen may conceivably have been purchased there, no tortoises are native to Puerto Rico, nor do Schwartz and Thomas (1975) nor Pritchard and Trebbau (1984) report even introduced tortoises on that island. The specimen itself is large — CL 57 cm — and on those grounds would appear to be referable to the larger of the two species, *Geochelone denticulata*, with *G. carbonaria* not being known to exceed 51.2 cm CL and usually much smaller (Pritchard and Trebbau, 1984). On the other hand, the uniformly black carapace with a single yellow spot on the areola of the second vertebral, is more typical of *G. carbonaria*, although in this species the yellow spots are usually present on each of the vertebrals and costals (but uniformly black specimens may occur in Bolivia).

One incidental comment of Garman's provides interesting insight into the slowness of communication between the Galápagos and the United States at the beginning of the century. Garman (p. 265) commented on the Cobos settlement on Chatham Island, and observed that it "may yet be in existence." In fact, this colony had come to an end thirteen years earlier, in 1904, when Cobos' own peons, tired of his tyrannical and violent rule, had murdered him with their machetes.

But despite the defects of Garman's work, it also had positive aspects. Some of Garman's taxonomic insight, although often discounted in recent publications, appears to be valid. Garman was in some cases able to discern the validity of species names (such as *Chelonia* (= *Natator*) *depressa* and *Testudo* (= *Geochelone*) *nigra*) that were ignored by his contemporaries and only restored to validity decades after his death. His discussion of the apparent changes in the island populations with time has similarly been recently vindicated. From the point of view of exploring the taxonomic history of the Galápagos tortoises, Garman's extensive series of plates are most valuable. And his conclusion that the type specimen of *Testudo ephippium* came from Abingdon Island, in agreement with Baur (1889), not Charles or Indefatigable as had been assumed by Günther (1877), nor Duncan Island as assumed by Van Denburgh (1914) and virtually all modern authors, seems to have been correct.

Current Nomenclature

Van Denburgh (1914) recognized 14 species of Galápagos tortoises (13 of them named), all of which he included in the genus *Testudo*. Van Denburgh had available to him the series of 256 specimens, of known island origin, recently collected by the *Academy* expedition, and he was thus able to describe and illustrate the various geographically isolated forms and to prepare a key for their identification. Van Denburgh recognized five species from Albemarle Island (*T. vicina*, *T. güntheri*, *T. microphyes*, *T. becki*, and the unnamed form from Volcan Alcedo), and one each from Narborough (*T. phantastica*), James (*T. darwini*), Jervis (*T. wallacei*), Indefatigable (*T. porteri*), Duncan (*T. ephippium*), Abingdon (*T. abingdoni*), Chatham (*T. chathamensis*), and Hood (*T. hoodensis*). The Charles Island species (*T. galapagoensis*) and the form from Barrington Island were found to be extinct. *Testudo phantastica* was based upon a single specimen (Fig. 20), fortunately adult; the immature specimen from Volcan Alcedo lacked distinctive characteristics (although unusually high-domed), and was wisely left unnamed. Van Denburgh found his single specimen from Jervis Island to be somewhat similar to Rothschild's type of *T. wallacei*, and he thus associated this name with the Jervis form even though the two specimens in question lacked any remarkable or obvious identifying features, and there was

Figure 20. Anterior and posterior views of holotype of *Testudo phantastica* (CAS 8101).

considerable suspicion that the one specimen known to have actually been found on Jervis had been introduced (by G. Baur).

Apart from the curious changes wrought by Garman (that found little acceptance), subsequently published modifications or additions to Van Denburgh's classification have been as follows:

i) De Sola (1930), having collected a series of live tortoises from the vicinity of Cartago Bay near Perry Isthmus, Albemarle Island (in the coastal lowlands between Volcan Alcedo and Volcan Sierra Negra), declared these to be referable to a new species, *Testudo vandenburghi*. Unfortunately, De Sola provided neither a diagnosis of this form nor a designation of the type specimen.

ii) Mertens and Wermuth (1955) reduced the various island forms to subspecies within the species *Testudo elephantopus*, and substituted the older name *T. nigrita* for *T. porteri*. Mertens and Wermuth considered *Testudo californiana* and *T. nigra*, both of Quoy and Gaimard, 1824 (but published in two papers, the one a few months before the other), to be questionable, and gave preference to the later name *T. elephantopus* Harlan, 1827. These authors also considered the five described forms from Albemarle (or six, if we include *T. macrophyes* Garman, 1917) as identical, listing all under *T. e. elephantopus*.

iii) Loveridge and Williams (1957) established *Geochelone* as the appropriate genus for the Galápagos (and many other) tortoises, the Galápagos forms being included (as *G. elephantopus*) with the South American mainland species in the subgenus *Chelonoidis* (rendered incorrectly as *Chelonoides* by Auffenberg, 1967 and 1971).

iv) Wermuth and Mertens (1961), in a revision of their inventory of the world's living turtle species, continued to use *Testudo* for the Galápagos tortoises.

v) In a further revision published after the death of Mertens, Wermuth and Mertens (1977) still used *Testudo*, but included the Galápagos tortoises in the subgenus *Chelonoidis*. Moreover, although in previous editions they had listed the various subspecifically indeterminate names for Galápagos tortoises as synonyms of *T. elephantopus*, in the 1977 edition these were all included in the synonymy of *T. e. elephantopus*. The copious literature between 1961 and 1977 listing the Galápagos tortoises as forms within the genus *Geochelone* was ignored even in the synonymies, and certain subspecies known to be living (those of Abingdon and Duncan islands), as well as the long-extinct Charles Island form, were all labelled as presumably extinct ("*vermutlich ausgestorben*"). No such label was attached to the one subspecies (*phantastica*) that may indeed be extinct, but that has not yet been proven to be so.

vi) Bour (1980), in a short paper based more on a philosophical striving for taxonomic balance within the family Testudinidae rather than new biological data, promoted the subgenus *Chelonoidis* to generic status, and with it elevated the various subspecies of Galápagos tortoises to full species status within that genus.

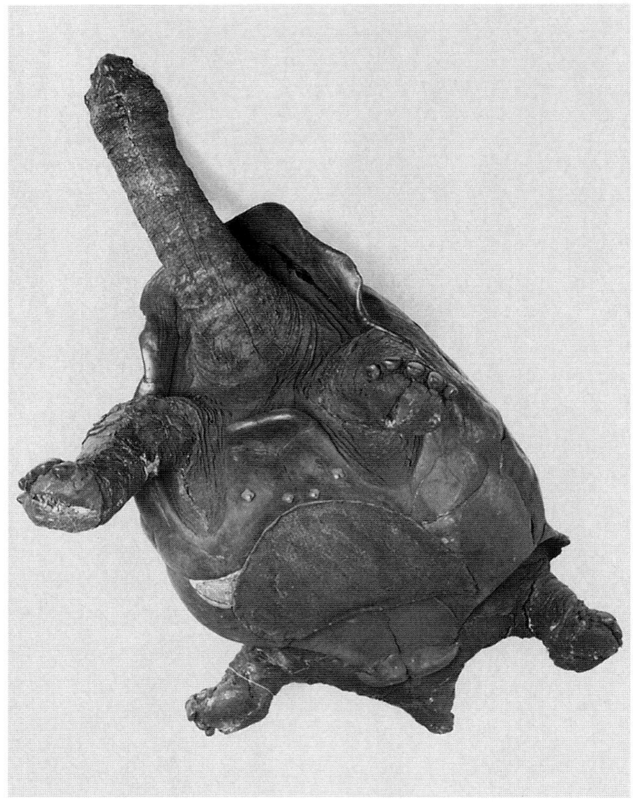

Figures 21 and 22. Dorsal and ventral views of the holotype of *Testudo ephippium* in the Royal Scottish Museum, Edinburgh. Photo courtesy A.S. Clarke, Royal Scottish Museum.

vii) Crumly (1982) considered the Galápagos tortoises to be subspecies of *Geochelone elephantopus*, but subsequently reversed himself (Crumly, 1984), arguing for full species status of each population, ironically in a paper demonstrating the virtual impossibility of distinguishing the island populations by skull analysis. Crumly's justification was that Fritts (1983, 1984) considered Van Denburgh's analysis, that used binomials, to be the authority of choice until superceded.

viii) Fritts (1983), in a study of the comparative morphology of Galápagos tortoises, largely (and probably wisely) refrained from providing scientific names for the different populations, with the notable exception of the caption to a figure in which the new combinations *Geochelone abingdoni*, *G. becki*, *G. darwini*, *G. epphipium* [sic], and *G. porteri* were perpetrated without comment. Further informal, perhaps even unrealized, new combinations (*Geochelone chathamensis*, *G. guntheri*, *G. vicina*, *G. hoodensis*, *G. microphyes*, *G. nigrita*, *G. wallacei*, *G. phantastica*, and *G. vandenburghi*) were published by Crumly (1984).

ix) Pritchard (1984), in a footnote to a paper on the Abingdon Island tortoise, observed that the type specimen of *T. ephippium* (Figs. 21, 22), as had previously been argued by Baur (1889), almost certainly came from Abingdon Island, not Duncan; and that the name *T. nigra* Quoy and Gaimard, 1824b, was published three years earlier than *T. elephantopus* Harlan, 1827, was not a *nomen oblitum*, and therefore was the valid specific epithet for the Galápagos tortoises (and had the incidental advantage of an extant holotype).

x) Bour (1985) partially reversed his earlier position that the Galápagos tortoises were all full species, considering them a complex of species and subspecies, and agreed with Pritchard (1984) that *nigra* was the appropriate specific epithet for this complex.

In view of this confusion, it is clear that a new analysis, from first principles, with reference to the latest edition of the Code of Zoological Nomenclature (ICZN, 1985), and using modern concepts of species and subspecies, is necessary to establish an appropriate nomenclature for the Galápagos tortoises. This should follow these steps: 1) Determination of the appropriate generic or subgeneric name for the group. 2) Determination of the biological relationship between the various populations. 3) Strict application of the ICZN Rules of Nomenclature to the biologically distinguishable populations. Elaboration of these three steps follows:

1) It must be observed that *genus* is a somewhat more subjective concept than *species*, and that no truly objective or biological determination may be possible. Mayr (1963) alludes to a number of island "monotypic genera" that turned out to be little (or nothing) more than subspecies of continental forms, and observed, on the other hand, that intraspecific variation in some vertebrate species may include such fundamental characters as number of cervical vertebrae or number of digits. Bour's (1980) proposal that the subgenus *Chelonoidis* be raised to generic rank was followed by

Pritchard (1984) primarily on the grounds that a proposed taxonomic change, if found unacceptable by another worker, should be refuted rather than ignored, a principle clarified by Lazell and Michener (1976). However, the Galápagos tortoises and their South American relatives lack the major shared derived features that distinguish the Seychelles/Aldabra/Madagascar tortoises (notably the extreme skull modifications), the Mascarene tortoises (single gular, hypertrophied odontoid tubercles, labial border of mandible strongly dentate, extensive shell and skull ankylosis — at least in the very old adults that constitute most of the available material), or the genera *Manouria* and *Indotestudo*, and recent authors such as Fritts (1983, 1984), Crumly (1985), and Pritchard and Trebbau (1984) have not accepted Bour's proposal, although Obst (1985) did.

The decision of Loveridge and Williams (1957) to list the Galápagos tortoises as members of the subgenus *Chelonoidis* within the genus *Geochelone* was not a unanimous one by these two authors. Williams prevailed, while Loveridge disclaimed in a lengthy footnote. Nevertheless, since both the revived genus *Geochelone* and the subgenus *Chelonoidis* were adequately characterized morphologically in this paper, the proposal should be considered a valid one even though the subgenus *Chelonoidis* was only found to differ from the subgenus *Geochelone* (from Africa and the Indian subcontinent) in two characters: the concealed postotic fenestra of the former and the narrowed pectoral scutes of the latter. In some Galápagos tortoises, the pectorals may be so narrowed as to fail to make midline contact; but tortoises of the subgenus *Geochelone* have pectorals that are not merely narrowed but that show a characteristic, parallel-sided form, although there is much variation and the character is a weak one at best. There is considerable ontogenetic influence in the development of narrowed pectorals, and the variation of pectoral narrowing within the extinct genus *Cheirogaster* is illustrated and discussed by Bourgat and Bour (1983). It appears to be associated with a phylogenetic shortening of the bridge and expansion of the humeral scutes.

Use of the subgenus *Chelonoidis* has been accepted by the majority of subsequent authors. Subgeneric names being optional, the publication of the binomial without the subgenus does not constitute disagreement with the original proposal. The conspicuous exceptions to this usage have been Wermuth and Mertens (1961, 1977) who persisted with *Testudo* for the Galápagos tortoises, without publishing a justification for their practice, and indeed, by failing to even list combinations involving *Geochelone* or *Chelonoidis* in their synonymies, seemingly declined to accept that these new allocations had even been proposed.

In my opinion, the arguments establishing *Geochelone* Fitzinger, 1835, as the appropriate genus for the Galápagos tortoises, and *Chelonoidis* Fitzinger, 1856, as the subgenus, are adequately documented and acceptable. The generic names *Chersine* Merrem, 1820, and *Testudinites* Weiss, 1830, both have chronological priority over *Geochelone*. However, *Chersine* is only in part a senior synonym of *Geochelone*, and Lindholm (1929) has designated *Testudo graeca* as its type species. Therefore *Chersine* (partim) as a synonym of *Geochelone* does not affect the stability of the latter. *Testudinites* Weiss, 1830, is definitely based upon South American material, but its suppression as a long unused name upon appeal to the International Commission on Zoological Nomenclature (ICZN) is assured.

If the Galápagos tortoises should be considered subgenerically (or generically) distinct from the South American mainland forms, a new name would have to be proposed. The name *Elephantopus*, used by Gray (1873) as a genus, has been proposed (Bour, 1988), but, as Bour points out, this name is a junior homonym of the coelenterate name *Elephantopus* Agassiz, 1846, and is therefore not available for the Galápagos tortoises.

2) Regarding the biological relationships between the various populations of Galápagos tortoises, the following points are relevant:

a) In no known case can tortoises of different Galápagos populations be morphologically distinguished as hatchlings,

Figure 23. Probably the largest tortoise in the world today, a male Indefatigable Island tortoise weighing over 350 kg at the Life Fellowship Ranch near Seffner, Florida. February 1993. Weight in October 1995 was 400 kg.

Figure 24. Juvenile tortoises (ca. 30 cm carapace length), from Hood Island (left) and from Cerro Azul, Albemarle Island (right), captive raised under identical conditions at the Charles Darwin Research Station. Note that the elevation of the anterior end of the carapace is already evident in the Hood Island tortoise, indicating genetic control of the saddlebacked trait. September 1982. Photo by A.G.J. Rhodin.

and in many cases (including the great majority of zoo specimens), the adults too cannot be attributed to a particular island of origin with any confidence. While Van Denburgh (1914) did publish a key by which the different island populations could be identified, this was in many cases based upon very small sample sizes, and in three cases (*phantastica*, *wallacei*, and the unnamed form from Volcan Alcedo) only single specimens were available.

b) Genetic studies by Marlow and Patton (1981) have revealed extremely close relationships between the various Galápagos tortoise forms.

c) Although Van Denburgh (1914) used binomials for the different populations, he was writing at a time when the subspecies concept was little used, and in his text he repeatedly referred to these populations as "races" rather than as "species." Yet those recent authors, such as Fritts (1983, 1984) who have utilized binomials cite Van Denburgh as the authority for this usage.

d) While it has been maintained by some that the failure of Galápagos tortoises to reproduce regularly or abundantly in zoos is a result of a mixed stock consisting of different "species" of tortoise (see e.g., Bacon, 1980), this argument ignores the fact that some of the zoo populations (especially at San Diego) were relatively large and composed of at most two or three island populations, so that even if cross-matings were unproductive, there would be ample opportunity for matings between members of a single population.

e) Ryder (1978) was unable to distinguish the populations of Galápagos tortoises by electrophoretic separation of serum proteins.

f) It is universally admitted that the Galápagos tortoises are more similar to each other than they are to any other species within (or outside) the subgenus *Chelonoidis*.

In sum, the above points are supportive of the hypothesis that the Galápagos tortoises are all referable to a single species.

Recognition of the various island or otherwise isolated populations as subspecies is also a point worthy of discussion. The populations are generally distinguished by adult size and the shape of the carapace (especially the anterior profile, the width at marginals 2–3, the scute or point at which the highest point is reached, and the flaring above the hind limbs). However, some other characters have also been used, including the amount of white or yellow coloration on the face and chin; whether the shell is black or brown, striated or smooth; or scute details such as the tendency in the Abingdon form for the upper edge of marginal 8 to be narrowed or even excluded from contact with costal 3, or for the pectorals to fail to meet on the midline of the plastron, as in some specimens from Chatham (Van Denburgh, 1914 and pers. obs.), from Duncan (van Lidth de Jeude, 1898; Van Denburgh, 1914), or in the single surviving individual from Abingdon (pers. obs.).

Some of these characters are better than others. Color and texture of the shell are closely involved with such features as age, sex, growth rate, and the overall humidity of the environment. The scute variations are too unstable to have major taxonomic value, and they appear to be primarily a superficial manifestation of changes in the overall shape of the shell (carapace and plastron). The extent of the unpigmented area of the face and throat, while generally more developed in the saddlebacked forms, definitely varies with age. Moreover, adult size is likely to be strongly influenced by feeding conditions. The largest tortoise reported in recent times is a captive raised specimen collected as a hatchling on Indefatigable Island as recently as 1957 (Fig. 23). It has been maintained on an abundant diet at the

Life Fellowship facility in central Florida, and now, although showing no signs of obesity, weighs about 400 kg (G. Moss, *pers. comm.*). Another feature, the degree of bossing of the carapace scutes, also appears to be related to speed of growth or to protein or calcium content of the diet, and is sometimes very marked in captive-grown animals — but reaches its greatest development in wild adult females of such dwarfed tortoise species as the South African tent tortoise, *Psammobates tentorius tentorius*.

Nevertheless, the development of the open shell front in the saddlebacked forms seems to be under genetic control, as the young tortoises of Duncan and Hood Islands, raised from hatching at the Charles Darwin Research Station under conditions identical to those under which young tortoises of dome-shelled races are raised, start to show the characteristic elevation of the nuchal area within a few years and at a carapace length of only about 30 cm (Fig. 24). On the other hand, those few individuals of the characteristically dome-shelled populations that display some degree of saddling of the carapace are invariably, as Fritts (1984) pointed out, "the largest and possibly oldest tortoises." Moreover, Marlow and Patton (1981) presented persuasive genetic evidence that the various saddlebacked populations had evolved independently from dome-shelled forms, in sharp contrast to Garman's unsupported conclusion that the forms *T. becki*, *T. ephippium*, and *T. hoodensis* were based upon young to medium-age specimens of *T. elephantopus*, and that *T. phantastica* and *T. abingdoni* were based upon old and very old specimens of the same species. These saddlebacked tortoises share a number of unusual, derived morphological features, and in the absence of trenchant differences, a modern cladist might also regard them as monophyletically derived from a dome-shelled ancestor.

Discussion of the "subspecies question" — the middle ground between considering the various populations as full species, or combining all into a single monotypic taxon — requires that consideration be taken of the changing concept of the subspecies among modern evolutionary biologists. In addition, comparisons and contrasts need to be made with the criteria that have been used by biologists evaluating variability in other chelonian superspecies, hybridizing species, intergrading subspecies, archipelagic taxa, *Rassenkreisen*, and the like.

The subspecies concept may have been first formalized in an editorial in Novitates Zoologicae (Anonymous, 1894), in which it was proposed that the term "variety" be discarded in favor of the two terms "aberration" (for individual variations) and "subspecies" for geographical forms which cannot rank as full species. Jordan (1896) elaborated: "A subspecies is a localized group of individuals of a species, the mean of the characters of which is different from the mean

Figure 25. Dead young adult male repatriated tortoise on Duncan Island, July 1986. This individual perished after the severe post-Niño drought of 1984–85.

of the characters of all other localized groups, and which will, under favorable circumstances, fuse together with other groups."

Working definitions of this type were utilized for much of the twentieth century, although the wording "under favorable circumstances" clearly needed refinement. The developing consensus was that subspecies generally intergraded where their areas of distribution joined, but it was also noted that captive conditions might promote hybridization between taxa that should still be considered as distinct species. Indeed, observations summarized by Karl et al. (1995) indicate that many if not most of the possible hybrids between different *genera* of cheloniid marine turtles may be produced, even under wild conditions, although all are very rare.

Wilson and Brown (1953) challenged the subspecies concept, raising a storm of protest from the "establishment" that rivalled the contrary outburst when subspecies were first proposed sixty years earlier. Wilson and Brown's argument centered upon the observation that so-called subspecific characters were frequently discordant — that is, that selection of different characters (for example, absolute size, or coloration, or relative head length) can result in completely different concepts of subspecific variation within a species. They concluded that the subspecies concept was arbitrary, and thus had no place in science.

This proposal gathered momentum in the ensuing years, and the tempo of descriptions of new trinomials faltered. But Wilson himself subsequently recognized that he and Brown had overstated their case (Wilson, 1994), and that, in actual fact, many species indeed showed geographic variation with genetic traits that changed in a concordant rather than discordant manner. In recent years, the general approach has been to utilize the criterion of intergradation for taxa whose natural ranges come into contact. From such cases may be developed a concept of the overall degree of morphological divergence that characterizes subspecies pairs as opposed to species pairs (that may typically exhibit widespread sympatry without intergradation), and to apply such a concept to determination of the status of allopatric populations. Nevertheless, this policy is not a unanimous one, and indeed Ernst and Barbour (1989) regularly utilize allopatry, rather than sympatry without intergradation, as a criterion for distinctness of chelonian taxa at the species level. But there is at least general agreement that a subspecies, in order to be valid, has to be geographically definable, and cannot be merely an assemblage of individuals found randomly throughout the range of a species.

The saddlebacked tortoises of Duncan, Hood, Abingdon, and Chatham are distinguishable morphologically. Van Denburgh (1914) translated these differences into quantitative criteria and percentages, but in general the tortoises of Hood and Duncan are small and have the highest point of the shell reached at the extreme front. The "saddle horn" of the Hood tortoise is narrowed and well-defined (Fig. 11); in the Duncan form it is broad and ill-defined (Fig. 25). The Abingdon tortoise — at least the adult males — is larger and has a much thicker neck and the shell reaches its highest point near the center of vertebral 2. The extant population of Chatham tortoises has not been described in any detail, but it is reminiscent of the Hood tortoise while less extreme — not as high, nor as narrowed, in front. The extinct form from Charles Island (see Figs. 44–45) was characterized morphologically by Broom (1929). I believe it is justified to regard these five forms as valid subspecies. Fritts (1984) also mentions the Barrington Island race, known only from bony fragments, as a saddlebacked race, but this is the only mention in the literature of the morphology of this population, and no details were given.

The subspecies situation is less clear on the central islands of Indefatigable, James, and Albemarle. Among the other reptiles on these islands, the populations of lava lizards (*Tropidurus albemarlensis*), geckoes (*Phyllodactylus galapagoensis*), and land iguanas (*Conolophus subcristatus*) are not considered distinct even at the subspecies level despite the existence of well-differentiated full species within these genera elsewhere in the archipelago. The habit of distinguishing the tortoises of the central islands as either species or subspecies may be largely a matter of habit, or a relict from the time when these populations were described as full species on the basis of individual type specimens with little or no supporting hypodigm or other information on population variability.

The concept of a different subspecies of tortoise on each of the five volcanoes of Albemarle Island is complicated by the fact that only on Volcan Alcedo are the tortoises most commonly found near the summit or in the caldera. Elsewhere they are more typical of the lowlands and intermediate altitudes (although they do reach the 5540 foot summit of Cerro Azul), and indeed populations have been reported in the lowlands equidistant from the mountains of Cerro Azul and Sierra Negra (near Cabo Rosa), between Sierra Negra and Volcan Alcedo (where De Sola collected his specimens of *T. vandenburghi* at Cartago Bay, in an area subsequently covered by new lava flows), and between Volcan Alcedo and Volcan Darwin (at Urvina Bay).

The shell color and texture of these tortoises, while subject to variation, may be environmentally determined (Fritts, 1983), as may be the adult size. The Volcan Darwin tortoise, in a rather xeric environment, is smaller than those of the three volcanoes to the south; A. Rhodin and I encountered a copulating male of this population with a carapace length of only 77 cm. There may thus be no adequate basis for distinguishing the Albemarle populations, except for that of Volcan Wolf, even at the subspecies level. The characteristic of the unusually eroded shells of adult tortoises on Sierra Negra, attributed by Van Denburgh (1914) to volcanic activity, may more reasonably be explained as a form of shell rot or fungus resulting from the unusually damp conditions under which the highland enclave of this population lived. Beck (1903) noted that, as one ascended Sierra Negra, the tortoises increasingly had eroded or scarred shells, considered by Fritts (1984) to be characteristic of the *guentheri* morphotype. Similar shell erosion is found in wild or introduced populations of the mainland species *Geochelone carbonaria* in very damp regions such as in the Darién of Panama or on Old Providence Island (pers. obs.). It is more of a testimony to the taxonomic microscope that has been directed at the Galápagos tortoises for more than a century than to any realistic subspecific characters that such environmentally derived features would even be considered to have nomenclatural value.

Nevertheless, recent writings by those most familiar with southern Albemarle tortoises in the field (e.g., Cayot and Louis, 1994) do follow Fritts (1984) in recognizing a *guentheri* morphotype in southern Albemarle, although they consider it to be far more restricted than did Van Denburgh (1914). Van Denburgh considered the *guentheri* morphotype (or race) to be distinguished by the relatively flat carapace, especially characteristic of large adult males, and to be distributed generally over the southern slopes of Sierra Negra ("Vilamil Mountain"), although it did not reach as far east as Cabo Rosa (where a form indistinguishable from *microphyes* occurred), and even within the stated range of guentheri there was a generous admixture of the *vicina* morphotype.

Picking through the ruins of these devastated populations today, Cayot and Louis (1994) reported that the *guentheri* morphotype still existed, but only in the vicinity of Cinco Cerros (i.e., southeastern slopes of Cerro Azul, not Sierra Negra — the supposedly exclusive range of *vicina*).

It is difficult to make sense of all of this. Of course, tortoises do move — very extensively when the terrain permits — and, over time (at least up to the time of the 1926 lava flow), tortoises may have wandered throughout the vegetated areas of southern Albemarle. In that the *guentheri* morphotype cannot be associated with any definitive geographic range, was reported by Van Denburgh to be microsympatric with specimens of the adjacent race *vicina*, and has characters that are really only expressed in adult males, it appears best to regard it as a synonym of *vicina*, pending the publication of DNA studies by Edward Louis.

The northernmost population of tortoises on Albemarle is a different matter. At least presently isolated from the Volcan Darwin population by lava flows, the specimens of this tortoise seen in the dry lowlands at Banks Bay and at Punta Albemarle include a significant proportion of strongly saddlebacked individuals (pers. obs.). Van Denburgh (1914) did not report dome-shelled tortoises on this mountain, but his photographs do demonstrate great variation in the degree of saddlebacking and anterior narrowing of the adult males, and the single adult female collected (CAS 8120) was, surprisingly, the most strongly saddlebacked tortoise found on Volcan Wolf even though it had a carapace length of only 21.75 inches. Rothschild (1901), in describing *T. becki* from northern Albemarle (on the basis of a single large male)

Figure 26. Very old female James Island tortoise, naturally imprisoned for many years in a steep-sided tuff crater south of James Bay. January 1970. The animal has since been liberated in the interior of James Island.

commented that this new form "was of the greatest interest, because it demonstrates the presence, not only of a third species of Giant Land Tortoise on Albemarle Island, but also a tortoise of a totally different kind from the other two species."

Yet nowadays, as I have seen at both Bank's Bay and Punta Albemarle, the Volcan Wolf population includes an apparent majority of tortoises that are somewhat flattened or dome-shelled, and not saddlebacked. Possibly a long-isolated population of saddlebacked tortoises, adapted to the arid and precipitous conditions of Volcan Wolf, was invaded by dome-shelled tortoises from the south during a period when the lava separating these mountains had become sufficiently eroded and vegetated as to be passable. If this postulate is correct, then the production of morphologically intermediate tortoises on Volcan Wolf today is further evidence that the populations are subspecifically related only, and do not show either mutual infertility nor behavioral reproductive isolating mechanisms. The persistence of extreme forms as well as intermediate ones suggests either that the populations met comparatively recently and have not yet become fully intergraded, or that relatively few genes control the saddling of the shell.

One could then add the Volcan Wolf tortoise to the list of valid subspecies of Galápagos tortoises on the basis of certain phenotypes that are definable and different from other saddlebacked tortoises in basically the ways described by Rothschild and Van Denburgh, although the population is so variable today that a rigorous diagnosis of the subspecies would probably be impossible. Moreover, it would still be appropriate to synonymize the other four subspecies reported from Albemarle; they simply cannot be distinguished with any certainty, and the most obvious characters (size, texture) are environmental responses.

The populations of James and Indefatigable are not very different from those of the southern volcanoes of Albemarle, but large males, at least, tend to be distinctive in minor ways, and certainly, being on separate islands, they may be considered genetically isolated. The James Island tortoise is a large, semi-saddlebacked or intermediate form (Figs. 15, 26), diagnosed by Van Denburgh as having the carapace width at marginals 2–3 equal to 48–58% of the straight carapace length, and the height at the nuchal notch 42–45% of the straight length. The younger adult males on Indefatigable have an extremely rounded, highly domed carapace, with the height at the nuchal notch 30 to 40% of the straight length, and the width at marginals 2–3 from 53 to 71% of the straight length. Very large, old males from Indefatigable (none of which were collected by Van Denburgh) may tend towards elongation of the shell, or posterior increase in width, producing a less prominently domed form (Figs. 16 and 27, also see photos on p. 6 in the editorial prologue and on the back cover). But the James and Indefatigable tortoises may be considered as valid subspecies on the grounds of geographic isolation and at least marginal morphological difference.

3) Under the ICZN Rules of Nomenclature, the oldest available name that is not a *nomen nudum*, whose holotype is clearly a representative of the species in question, and that was published in a work exclusively employing the binomial system, is the one that should apply. Formerly, long-forgotten senior synonyms could be dismissed as *nomina oblita*, but the latest edition of the Rules (ICZN, 1985) requires formal petition be made before such names can be considered unavailable. *Testudo rotunda* Latreille, 1801, although thought to apply to a Galápagos tortoise by Bour (1980), has subsequently been shown by the same author to apply to a Mascarene species (R. Bour, *in litt.*), so the first name to be applied to a Galápagos tortoise in the scientific literature is *Testudo californiana* Quoy and Gaimard, 1824a.

This description, consisting of a brief Latin diagnosis ("*Testudo toto corpore nigro*; *testa gibba*; *scutellis dorsalibus priori posteriorque elevatis*; *loricae cunctis margine striatis*; *lateribus subcarinatis*") was followed by a series of dimensions, using the ancient system of *pieds* or *pedes*, *pouces* or *polles*, and *lignes* or *lines*. This is a duodecimal system, with 1 *pouce* equivalent to 27 mm. The paper announcing this new species was read at the meeting of the Société d'Histoire Naturelle de Paris on 7 November 1823 (J. Lescure, *pers. comm.*), by M. Gaimard.

The proceedings of this meeting were published in the second section of the Bulletin Universelle des Sciences et de l'Industrie, part of the Bulletin des Sciences naturelles et Géologie. The precise date of this publication is unclear, but I am advised by L. Jescure (*in litt.*, 12 March 1977) that it was very probably before 24 April 1824, and certainly before 18 September 1824.

The date is important, in that, later in 1824, a new description of the same tortoise was published, also by Quoy and Gaimard (1824b), within the work *Voyage Autour de Monde executé sur l'Uranie et la Physicienne pendant les années 1817–1820* by M.L. de Freycinet. This was published in fascicles; Fascicle 4, containing the description of the new tortoise, this time called *Testudo nigra*, appeared on 18 September 1824, with the plate (Pl. 40) published later (18 December). The new description (differing by only a few words from the earlier description of *T. californiana*) was supplemented by a page and a half of French text describing the specimen in fair detail, and giving the supposed origin as California.

It is curious that, after all of the mariners' and buccaneers' accounts of giant tortoises in the seventeenth and eighteenth centuries, it seems to have been still unrecognized in the scientific world that giant tortoises even existed — in the paper immediately following the description of *T. californiana*, Fitzinger (1824) gives an account of his visit to the Menagerie in Vienna, where he identified a Brazilian tortoise that he called *Testudo schweigeri* (presumably the form now known as *Geochelone denticulata*, a large but not giant species) as the largest known land tortoise ("*la plus grande tortue de terre que l'on connaisse*"). Schweigger's description of this species as *T. gigantea* is discussed by Pritchard (1986). But the name *T. schweigeri* ("corrected" by Boulenger [1889] to *Testudo schweiggeri*), is normally

Figure 27. Adult male tortoise in lush vegetation near the *Caseta*, southeastern highlands of Indefatigable Island. July 1986.

attributed (e.g., by Boulenger [1889] and Wermuth and Mertens [1961]) to a later writer than Fitzinger (1824), i.e., Gray (1831), and based upon a rather small shell so puzzling in appearance that no one has been able to identify it. (It may be based upon a specimen of *Gopherus polyphemus* showing numerous abnormalities of both form and coloration). But sometimes the task of making precise interpretations from decidely casual writings by nineteenth century naturalists, who all too frequently took no note of each others' work, leads one to the type of conclusion expressed by Vanzolini (1977), who commented that "Gray's carelessness in citing scientific and personal names, his disregard for other people's opinions and previous usage, and wretched proof reading, all conspire together to make this pretentious little work [i.e., Gray, 1825] very unpleasant."

Testudo californiana Quoy and Gaimard, 1824a, was not utilized by subsequent reviewers. Under earlier editions of the Rules of Nomenclature, this name could be dismissed, either because it was a *nomen oblitum* or because the description was an unrigorous written summary of a verbal presentation. Under the 1985 Rules, however, these considerations no longer apply, and the name will have to be used unless succesful petition is made to the ICZN to have it repressed. On the other hand, even under the old Rules, *Testudo nigra* could not be dismissed as a *nomen oblitum*. A name used by Charles Darwin (1845) in his *Journal of Researches* is hardly a forgotten one, and Garman (1917) used this name for the tortoise of Charles Island. It was also cited by Cuvier (1829), Duméril and Bibron (1835), Wiegmann (1835), Strauch (1862, 1865, 1890), and Boulenger (1889), and was discussed by Van Denburgh (1914) and by Rothschild and Günther (in Rothschild, 1902c), although these last authors, convinced that the island populations should all be referred to different species, dismissed the name as unusable because the holotype was too small for the island of origin to be determined. Nevertheless, this problem is only operative as long as each population is considered a distinct species. When Mertens and Wermuth (1955) subsumed all the Galápagos tortoises within a single species, they should have selected *nigra* as the valid epithet with priority, and dealt separately with the issue of which island population should be considered the nominate subspecies.

The species name *elephantopus* of Harlan (1827) has, of course, been used in a vast number of papers during the last thirty years, and a case could be made for a petition to the ICZN to retain it. However, I do not recommend this step, in part because use of *elephantopus* for all the Galápagos tortoises is relatively new (since 1955), and thus lacks the status of a *nomen veneratum* as defined by Smith and Smith (1980), and in part because the island of origin of the holotype is conjectural. Van Denburgh's assumptions leading to his guess that this might be the Charles Island form were clearly flawed (Broom, 1929). Moreover, the holotype, being lost (the appalling confusion resulting from efforts to find or identify this specimen being discussed by Garman, 1917), is not available for future, more sophisticated morphological, genetic, or biochemical analysis that might have clinched the identification. I therefore conclude that the correct specific name for the Galápagos tortoises is *Geochelone* (*Chelonoidis*) *nigra* (Quoy and Gaimard, 1824b).

At the subspecies level, further nomenclatural problems exist. If *nigra* is the applicable epithet for the species, which race is the nominotypical subspecies, *Geochelone nigra nigra*? Until such time as Captain Meek's log books are found and studied and the actual origin of the type specimen determined (an improbable development), one option would be to make an arbitrary determination, validated in ways dictated by a careful consideration of the Rules of Nomenclature. However, perhaps the least painful option would be to turn a deliberate blind eye to the weaknesses in Garman's arguments that this form was from Charles Island, and to accept that designation on the grounds that it at least cannot readily be disproved. The procedure would have the advantage of not affecting the nomenclature of any extant subspecies.

Furthermore, only three of the subspecies of Galápagos tortoises (*hoodensis*, *phantastica*, and *darwini*) seem to have unclouded taxonomic status, with demonstrable morphological distinctness, known islands of origin of the type material, and no known senior synonyms (Van Denburgh, 1907), although *T. phantastica* Van Denburgh, 1914 does represent a minor correction of the original name *T. phantasticus* Van Denburgh, 1907. The holotypes of these three forms are, respectively, CAS 8121, CAS 8101, and CAS 8108. Problems with the other names are as follows:

i) While *T. chathamensis* was based by Van Denburgh (1907) upon a holotype (CAS 8127, Fig. 17) definitely collected on Chatham Island, the form described from there (based on a single entire specimen and a few shells and bones of tortoises that had died naturally or been killed by settlers) was a flat-shelled type found in the relatively humid areas southwest of the fresh lava flows of central Chatham. This population was almost extinct in 1906, when Van Denburgh's specimen was collected, and it now appears to be completely so. The photograph of Chatham Island tortoises in Sulloway (1984), with the caption indicating that they were located in northwestern Chatham Island, is misleading; the caption incorporates a typographical error and should read north*eastern* Chatham Island (F. Sulloway, *pers. comm.*).

The extant population of tortoises on Chatham, mentioned by Dorst (1965), Castro (1970), and Pritchard (1979), and numbering several hundred individuals, is rather markedly saddlebacked (at least in the adult males), and occurs in the dry lowlands northeast of the barren volcanic center of Chatham. It thus seems that the living population is not true *chathamensis*, but rather an undescribed subspecies.

ii) The type specimen of *T. becki* (No. 2 in the Rothschild collection, an adult originally described as 40.75 inches = 103.0 cm in length [Rothschild, 1901]; clarified by Rothschild [1915a] as being *curved* curved length, and with a straight carapace length of 31.25 inches = 79.4 cm), was reportedly not from Volcan Wolf, to which population this name is customarily applied, but from Cape Berkeley (Volcan Ecuador), at the northwestern tip of Albemarle, and part of a separate volcano (now bisected by downfaulting). It is unclear if this is simply an error of labelling, or if the specimen really came from Cape Berkeley. It could also be a case of geographic confusion. The original description referred to "Cape Berkeley, northern point of Albemarle island" (Rothschild, 1901), although the northern point of Albemarle Island is in fact Punta Albemarle, a known habitat for saddleback tortoises. Cape Berkeley is one of the least-known areas of the archipelago today, and while no tortoises have been found there in recent years, the possibility that some are there cannot be dismissed. Conceivably there is (or was) a population of saddlebacks on Cape Berkeley that intermittently had access to Volcan Wolf, and the tortoise population on the latter mountain may represent an intergradient one between saddlebacked and domed tortoises. But this is speculation.

iii) The form *T. wallacei* was described by Rothschild

Figure 28. Illustration from Snow (1964) of the male saddlebacked tortoise found by the Angermeyer brothers in northwestern Indefatigable Island in 1962. Photo by D.W. Snow. The original caption read: "The old male saddle-backed tortoise on Indefatigable feeding on a cactus pad from the opuntia tree behind him. The devastation of the undergrowth is mainly the work of goats."

Figure 29. Juvenile tortoise found by the author near Saddleback Hill (Cerro Montura), Indefatigable Island, in May 1972. The small size of the animal (ca. 35 cm CL) is suggestive of successful reproduction of this colony since its discovery in 1962.

Figure 30. Reasonably fresh tortoise droppings found near Saddleback Hill by A.G.J. Rhodin and the author in September 1982.

(1902d) on the basis of a single smooth-scuted adult male shell 81.9 cm in straight length, No. 42 in the Rothschild collection, whose origin was unknown. Originally labelled with the catchall giant tortoise name *T. indica*, it came from the private collection of one Mr. Bullock. It was then acquired by the Wallace collection, housed in Distington, Cumberland, and passed into the Rothschild collection upon the dispersal of Wallace's collection. Rothschild speculated that the specimen was from Chatham Island, on the thin argument that, between 1800 and 1835, most of the tortoises collected in the Galápagos came from James and Chatham Islands, and this specimen, not being as rounded as the James tortoises were known to be, presumably came from Chatham.

Van Denburgh (1914) reported on the collection of a single live male tortoise on Jervis Island by Beck and Slevin, which he found to be generally similar to the type of *T. wallacei*. He thus felt tentatively justified in using this name for the tortoises of Jervis Island, even though it was profoundly doubtful that Jervis had ever supported a native tortoise population. Van Denburgh did concede that there was a rumor in the islands that Baur had introduced tortoises to Jervis, and Rothschild (1915a) considered the differences between the specimens to be rather profound. In view of these circumstances, the name *T. wallacei* Rothschild 1902 should be considered a synonym of *T. nigra* Quoy and Gaimard, 1824b.

Nevertheless, the conclusion of Pritchard (1979) and others that Jervis Island never had a native tortoise population must be re-examined in the light of reported discoveries of fossil or cave remains of both tortoises and land iguanas on this island, that led Steadman and Zousmer (1988) to state that Jervis once supported native populations of tortoises and land iguanas that might have died out as a result of predation or increased competition for food sources. Their speculation is questionable, in that the rats and (formerly) goats on Jervis were both of recent introduction, and could have played no part in the extermination of terrestrial reptiles. Perhaps the tortoise bones found derived from an unsuccessful introduced population; G. Baur, for one, is known to have liberated tortoises on Jervis.

iv) The names *porteri* and *nigrita* have been used virtually interchangeably for the tortoises of Indefatigable Island. *Testudo nigrita* Duméril and Bibron, 1835, is the older name, but is based upon a very young and a 55.9 cm subadult specimen, the former in the MNHNP collection, the latter (on which the description was primarily based, and thus to be considered the lectotype) in the Hunterian Museum (Royal College of Surgeons). Günther (1877) later associated a large (104.1 cm), domed carapace in the Royal College of Surgeons Collection (now in the BMNH) with this name, and Garman (1917) figured this specimen in his Plate 9. Van Denburgh (1914) wrote that "we seem justified... in saying that Günther's *Testudo nigrita* agrees with the Indefatigable tortoise," but still preferred the use of the name *T. porteri* Rothschild, 1903, whose holotype, No. 44 in the Rothschild Museum, was a 92.7 cm adult male definitely collected on Indefatigable Island (by R. Beck). Since *porteri* has achieved rather widespread use (e.g., by Van Denburgh, 1914; Pritchard, 1967; Bailey, 1970; Groombridge, 1982; and Fritts, 1983), its continued use is justified, especially in view of its more satisfactory type material. While the large Royal College of Surgeons specimen, identified as *nigrita* by Günther (1877), is almost certainly an Indefatigable tortoise on grounds of morphology, it must be remembered that this individual was not part of the type series, and it would thus be appropriate to make petition to the ICZN to suppress the name *nigrita*.

There is another isolated and very small population of tortoises on Indefatigable, described by Snow (1964) and by Pritchard (1979) (Figs. 28–30). Adult tortoises of this population are strongly saddlebacked; they live in the arid northwestern corner of the island, separated by a considerable distance from the population of *porteri* in southwestern Indefatigable. This population has not been named, and there are no specimens in museum collections.

v) The name *T. abingdoni* Günther, 1877, is customarily used for the tortoises of Abingdon Island, and indeed this name is based upon an adequate series of specimens (BMNH 76.6.21.38–40, 76.6.21.44) of definite Abingdon Island origin (collected by J. Cookson in 1875) (Fig. 31, see also frontispiece painting). However, as I have previously argued (Pritchard, 1984), the holotype of *T. ephippium* Günther, 1875, customarily allocated to Duncan Island by Van Denburgh (1914) and subsequent authors, agrees in both morphology and circumstantial details of its collection with the tortoises of Abingdon (Baur, 1889), and is the older name. The holotype of *ephippium* (Figs. 21–22), although a relatively young animal, has a carapace length of 83.8 cm — about 9 cm longer than the largest of 86 Duncan tortoises collected by the *Academy* expedition (Van Denburgh, 1914), and about 7 cm longer than the largest of 26 specimens in the Rothschild collection (Rothschild, 1915a), but absolutely typical of male Abingdon tortoises (Rothschild, 1915a). Moreover, in lateral profile, the specimen, still in good shape in Edinburgh, shows the Abingdon tortoise characteristic (contrasting with the Duncan tortoise morphology) of the highest point of the carapace reached at vertebral 2 rather than at the very front of the carapace.

The circumstances of the collection of the specimen are also persuasive. Baur (1889), referring to the diary of Captain Basil Hall (1840), a mariner who visited the Galápagos in 1822, determined that Abingdon was the only island that Hall had actually visited in the archipelago, and that he had taken some tortoises on board, although all perished in the cold weather off Cape Horn. One of these was preserved, and "it may now be seen in the Museum of the College at Edinburgh." This collection (now the Edinburgh Museum of Science and Arts) did not give catalog numbers to specimens in the early nineteenth century, but the acquisition log does include an entry indicating "Large turtle from South Sea — Captain Basil Hall," and it only has a single Galápagos tortoise — the type of *T. ephippium*.

Thus, *T. abingdoni* Günther, 1877, is a junior synonym of *T. ephippium* Günther, 1875, and the Duncan Island tortoise will require an alternative name; the discovery that a type specimen has been misidentified is a more serious problem than the mere discovery of an obscure senior synonym, which can often be disposed of by petition to the ICZN. It is appropriate to revive Garman's *Testudo duncanensis* for this purpose. This name appears only once, on p. 269, of Garman (1917), where it is offered in binomial form, although described as only a "variety" of *Testudo elephantopus*. Later in this paper (pp. 290–296), Garman offers a detailed description of his composite taxon *T. elephantopus*, and on pp. 292–293, there appears a detailed description of a 25 inch male specimen, MCZ 11068, with

Figure 31. Mounted specimen of an Abingdon Island tortoise, a very old adult male collected by J. Cookson in 1875; it died off Cape Horn on the journey to England. Specimen in the British Museum (Natural History). See also the frontispiece painting which appears to be based on this specimen.

accurate collection data — it was collected on Duncan Island by G. Baur. The account also describes adult female specimens from the same island. Thus, MCZ 11068 should be considered the lectotype of the Duncan Island tortoise, *Geochelone nigra duncanensis* (Garman, 1917), and is so designated.

On the other hand, substitution of the name *ephippium* for *abingdoni* would cause extensive confusion, the former having a substantial recent literature in all of which it is considered to refer to the Duncan tortoise, and it is appropriate to petition the ICZN to establish the validity of *abingdoni* by Plenary Decree despite the existence of a senior subjective synonym.

vi) While this point becomes moot if the tortoises of the four southern volcanoes of Albemarle are synonymized, Garman observed that the name *T. microphyes* Günther, 1875, customarily used to this day for the tortoises of Tagus Cove and Volcan Darwin, was based upon a single, probably abnormal female (but ventrally concave) specimen, 57 cm in length, of unknown provenance. It had been purchased by the British Museum from the Museum Committee of the Royal Institution of Liverpool. Günther's subsequent allocation to *T. microphyes* of a series of specimens of known Volcan Darwin origin (from "a small elevated plateau covered with stunted bush and high, very coarse grass" about four miles inland from Tagus Cove, where tortoises may still be found during moist times of the year) did not exempt the holotype from challenge, and in view of this, Garman (1917) proposed the alternative name *T. macrophyes* for the Tagus Cove tortoises, although for mysterious reasons he continued to use *T. microphyes* for other Albemarle populations.

Another detail concerning the nomenclature of one of the southern Albemarle tortoises pertains to the form described by Baur (1889) as *T. güntheri*, written thus, i.e., with an *umlaut*. Many authors from then to now have rendered this name in identical fashion — these include Van Denburgh (1914), Rothschild (1915a), Garman (1917), Beebe (1925), Eibl-Eibesfeldt (1960), Wermuth and Mertens (1961), Hendrickson (1964), Pritchard (1967), IUCN (1968), Black (1973), MacFarland et al. (1974a, 1974b), and de Vries (1984). Crumly (1984) both utilized and quoted the epithet without the *umlaut*; however, a corrigendum inserted into this paper indicated that *güntheri* should be substituted for *guntheri* throughout.

Despite the abundance of authors who used the *güntheri* form, the 1961, 1964, and 1985 editions of the Rules of Nomenclature (as well as the draft version of the forthcoming edition) are quite explicit; in Chapter VII, Article 32 (d) (i) (2), they specify that a name derived from the German and including a "ü" should be emended so that "ue" is substituted for "ü." On the other hand, the 1977 Rules included additional wording, specifically: "...*except* when the name was first corrected by deletion of the mark concerned, in which case it cannot be corrected further."

Authors who used *guentheri* include Honegger (1972, 1980), IUCN (1975, 1979), Corley Smith (1977), Groombridge (1982), Fritts (1983, 1984), Obst (1985), and Wermuth and Mertens (1977). These last authors considered this taxon invalid, listing it in the synonymy of *T. e. elephantopus*, misquoting the original author (Baur, 1889) by using the emended form (items in synonymy should be cited exactly as originally written). However, the earliest use of this emended form appears to have been by me (Pritchard, 1971a, 1971b) in two papers in which, to my embarrassment, I used first the unaccented (*guntheri*) form and then the emended form (*guentheri*). I found to my further chagrin that in my 1967 book, the accented form (*güntheri*) had been used in the text, while in the appendix to the same volume it had appeared in unaccented form (*guntheri*). To date, Bailey (1970), Iverson (1985) and Cayot and Louis (1995) are among the few who have followed me in the usage of the unaccented form.

Further problems are encountered with the name of the adjacent population, *T. vandenburghi* De Sola, 1930. This name, a *nomen nudum*, can only be validated if a type specimen is declared and if the features by which the form differs from other Galápagos tortoises are clarified. Moreover, the name can only be legally applied to the Volcan Alcedo tortoise if it can be shown not only that the population from which De Sola's specimens were obtained (a lowland population near Cartago Bay) was identical to the Alcedo population, but also that it differed from other Albemarle populations with older names. This demonstration will be difficult, in that the Cartago Bay population appears to have been eliminated from the wild by volcanic activity, but perhaps it would be possible to trace some of the specimens De Sola sent to zoological parks, if they happened to end up in museums upon their demise. However, if they survived for any length of time in captivity, they are likely to have developed some minor abnormalities of captive growth that could mask any subtle shell features unique to this population. Interestingly, even though the Volcan Alcedo tortoise population is the most abundant in the archipelago today, it has never been morphologically characterized in the years since De Sola's 1930 publication, and Van Denburgh's (1914) original designation of the term *Testudo* sp. for this population relied entirely upon a single specimen (CAS 8141) that was not only too small (carapace length 26.75 inches) to demonstrate potential subspecific characters, but was also atypically high domed. The "seventy old skeletons" observed by Rollo Beck on Volcan Alcedo when the live juvenile was found would have been useful, but unfortunately were not collected.

Further questions arise in connection with the selection of the appropriate trinomial for the single subspecies of tortoise that, I have argued, inhabits the four southern volcanoes of Albemarle. Of the available names (*T. vicina*, *T. elephantopus*, *T. microphyes*, *T. macrophyes*, *T. guentheri*, and *T. vandenburghi*), it should be noted that the origin of the type specimens of the first three forms and of *T. guentheri* are unknown; *T. macrophyes* has never been used since its original proposal; and *T. vandenburghi* remains a *nomen nudum*. However, the holotype of *T. vicina* Günther, 1875 (an adult male skeleton, BMNH 74.7.15.1, obtained from

Figure 32. Very old adult male tortoise on Volcan Alcedo, Albemarle Island, with atypically elongate, saddlebacked carapace. Desquamating carapacial texture is probably a result of protracted exposure to extremely moist conditions. March 1975.

Professor Huxley), was subjected to detailed morphological analysis by Van Denburgh (1914) and determined by him to be allocatable to the Iguana Cove (southwestern Albemarle) population. *Testudo microphyes* Günther, 1875, on the other hand, although having page priority over *T. vicina* (it was described in the same publication) was based on a specimen too small to be allocatable to a particular population on the grounds of morphology.

Testudo güntheri was proposed by Baur (1889), for a probably very heterogeneous series of five specimens that Günther (1877) had allocated to *T. elephantopus*. Baur considered that *T. elephantopus* and *T. vicina* were synonymous, and, in that Günther's series was not (in Baur's opinion) identical to *T. vicina*, it needed a new name. The arguments both in favor of and against this allocation were far from persuasive, especially in view of the fact that Baur did not formally declare a holotype for his *T. güntheri* — he was really only renaming Günther's series, and Günther was not using a new name, but rather allocating new specimens to Harlan's 1827 name. The largest specimen in Günther's series was a 78 cm old male skeleton in the Oxford University Museum (OUM 8656), of the flat-shelled type similar to Van Denburgh's males from Sierra Negra. The remaining specimens (an immature male and female, and two juveniles, from the Free Public Museum in Liverpool, the British Museum, and the Royal College of Surgeons Museum) serve to confuse rather than clarify, in that none has known provenance nor obvious distinguishing features, yet all must be considered part of the type series, with the actual holotype undeclared. The only specimen large enough to display the depressed carapace considered diagnostic of the *guentheri* subspecies or morphotype is the Oxford University Museum specimen, described by Günther (1877) as: "An adult male example: a perfect skeleton with carapace, but without epidermoid plates. The carapace is 31 inches long. History of the specimen unknown; purchased of a dealer in Paris for the Oxford University Museum, and kindly lent to me by Professor Rolleston, F.R.S." Although I know of no literature in which this specimen is identified as the holotype of *T. guentheri*, it is so identified in the Oxford University Museum catalogue (I. Swingland, *in litt.*, 29 October 1974). I therefore hereby formalize what has already been assumed, and declare this specimen (OUM 8656) the lectotype of *Testudo guentheri* Baur, 1889. As a parenthetical footnote, Rothschild (1902b), after examining the alleged holotype of *T. elephantopus*, decided that there were no grounds for Baur to have proposed a substitute name for Günther's series.

It thus seems appropriate here, as first reviser, to utilize the epithet *vicina* over the simultaneously published *microphyes* for the subspecies inhabiting both the high and low terrain of the four southern volcanoes of Albemarle Island. While some minor morphological divergence may exist among these populations (e.g., *guentheri* shows a tendency for large adult males to be somewhat flat-shelled), the proportion of specimens in which distinctive characteristics are present is too low to justify nomenclatural recognition of different subspecies. A few individuals of all of these populations may show a tendency towards markedly saddlebacked shells (e.g., a large male I saw alive on Volcan Alcedo [Fig. 32]; or the southeastern Albemarle specimen figured by Garman [1917] in his Plate 16), but these are too rare to affect the overall designation of these populations as basically dome-shelled. They do, however, serve to indicate that the genes for the saddlebacked shell may be present in all populations in a certain frequency, and that when such phenotypes are favored by appropriate environmental conditions they may be manifested throughout the population relatively rapidly. But apart from selection for a saddleback shell morphology on the drier, more barren islands, most of the other features that have been utilized to differentiate subspecies or populations — overall size, shell smoothness or sculpturing and degree of erosion, degree of bossing of the carapace scutes, and details of shell proportions — are surely responses of the individual to environmental circumstances rather than genetic differences. Moreover, under primordial conditions, Galápagos tortoises were subject to no predation once they had passed the very early growth stages, and as the only large herbivores in their environment, there is no reason to believe that fitness would be less in tortoises demonstrating minor divergences from the typical shell form, as long as the architectural, supportive, and thermoregulatory functions of the shell and shell openings were not compromised, and mating success was not reduced.

The large size and greater tendency towards shell saddling of the adult males of most or all Galápagos races may thus be manifestations of sexual selection, and the requirements of enhanced mating success. Had adult males not been available to taxonomists, it is probable that 80% of the names proposed for "new" Galápagos tortoise species would never have seen the light of day.

Summary of Proposed Nomenclature

The Galápagos tortoises are included in the genus *Geochelone* and the subgenus *Chelonoidis*. Technically, the specific epithet should be *californiana* Quoy and Gaimard, 1824a, but until this name can be petitioned to be suppressed by the ICZN, in the interim, as a convenience and to avoid giving credence to a name that I trust has no future, I shall utilize the epithet *nigra*, a name originally proposed as *Testudo nigra* Quoy and Gaimard, 1824b, as the combination *Geochelone (Chelonoidis) nigra*.

Synonymy of *Geochelone (Chelonoidis) nigra*

1824a *Testudo californiana* Quoy and Gaimard
Bull. Sci. Nat. Paris 1:90. Type locality: "Californie." (An unused name to be suppressed by petition to ICZN).

1824b *Testudo nigra* Quoy and Gaimard
Voy. Uran. Phys. Zool., 174. Type locality: "Californie."

1827 *Testudo elephantopus* Harlan
J. Acad. Nat. Sci. Philadelphia 5(2):284. Type locality: "Galapagos Islands." *Nomen dubium* (at subspecific level).

1835 *Testudo nigrita* Duméril and Bibron
Erpétol. Gen. 2:80. Type locality: unknown. *Nomen dubium* (at subspecific level).

1854 *Testudo planiceps* Gray
Proc. Zool. Soc. London 1853:12. Type locality: "Galapagos Islands." *Nomen dubium* (at subspecific level).

1875 *Testudo microphyes* Günther
Phil. Trans. Roy. Soc. London 165:275. Type locality: "Hood's Island" (by supposition). *Nomen dubium* (at subspecific level).

1902 *Testudo wallacei* Rothschild
Novit. Zool. London 9:619. Type locality: "Chatham Island?" *Nomen dubium* (at subspecific level).

1917 *Testudo clivosa* Garman
Mem. Mus. Comp. Zool. 30:283. Type locality: "Mascarenes?" *Nomen dubium* (at subspecific level).

1917 *Testudo typica* Garman
Mem. Mus. Comp. Zool. 30:285. Type locality: unknown. *Nomen dubium* (at subspecific level).

1952 *Testudo (Chelonoidis) elephantopus*, Williams
Bull. Amer. Mus. Nat. Hist. 99:555.

1967 *Geochelone (Chelonoidis) elephantopus*, Pritchard
Liv. Turt. World, 156.

1980 *Chelonoidis elephantopus*, Bour
Bull. Mus. Nat. Hist. Nat. Paris (4)2: 546.

I recognize ten subspecies of *Geochelone nigra*, with their referenced maps, photographs, type localities, synonymized names, and islands or areas of origin, listed as follows:

Geochelone nigra nigra (Quoy and Gaimard, 1824b)
(Maps 1, 8; Figs. 6–8, 44–45)
Type locality: "Californie." Restricted to Charles Island (Santa María or Floreana) (extinct).
Testudo nigra Quoy and Gaimard, 1824b.
Testudo galapagoensis Baur, 1889.
Type locality: Charles Island.
Testudo elephantopus galapagoensis, Mertens and Wermuth, 1955.
Chelonoidis galapagoensis, Bour, 1980.

Geochelone nigra abingdoni (Günther, 1877)
(Maps 1, 2; Figs. 21–22, 31, 34–35, frontispiece)
Type locality: "Abingdon Island" (Pinta) (extinct in the wild; one captive survivor). Name requires validation by the ICZN by suppression of the senior synonym *Testudo ephippium* Günther, 1875.

Testudo ephippium Günther, 1875.
Type locality: Charles Island (in error). Name previously used erroneously for the Duncan Island subspecies.
Testudo abingdoni Günther, 1877.
Type locality: Abingdon Island.
Testudo elephantopus ephippium, Mertens and Wermuth, 1955.
Chelonoidis abingdoni, Bour, 1980.
Chelonoidis ephippium, Bour, 1980.
Geochelone abingdoni, Fritts, 1983.
Geochelone epphippium, Fritts, 1983.

***Geochelone nigra becki* (Rothschild, 1901)**
(Maps 1, 10; Fig. 19)
Type locality: "Cape Berkeley, northern point of Albemarle Island." Northern and western slopes of Volcan Wolf, Albemarle Island (Isabela).
Testudo becki Rothschild, 1901.
Geochelone elephantopus becki, Pritchard, 1967.
Chelonoidis becki, Bour, 1980.
Geochelone becki, Fritts, 1983.

***Geochelone nigra chathamensis* (Van Denburgh, 1907)**
(Maps 1, 5; Fig. 17)
Type locality: "Chatham Island." Southwestern and central Chatham Island (San Cristóbal) (extinct).
Testudo chathamensis Van Denburgh, 1907.
Testudo elephantopus chathamensis, Mertens and Wermuth, 1955.
Geochelone elephantopus chathamensis, Pritchard, 1967.
Chelonoidis chathamensis, Bour, 1980.
Geochelone chathamensis, Crumly, 1984.

***Geochelone nigra darwini* (Van Denburgh, 1907)**
(Maps 1, 9; Figs. 14, 15, 26)
Type locality: "James Island" (San Salvador or Santiago).
Testudo darwini Van Denburgh, 1907.
Testudo elephantopus darwini, Mertens and Wermuth, 1955.
Geochelone elephantopus darwini, Pritchard, 1967.
Chelonoidis darwini, Bour, 1980.
Geochelone darwini, Fritts, 1983.

***Geochelone nigra duncanensis* (Garman, 1917)**
(Maps 1, 3; Figs. 1, 2, 10, 25, 36–41, front cover)
Type locality: "Duncan Island" (Pinzón). This subspecies previously erroneously designated *Testudo ephippium* Günther, 1875.
Testudo duncanensis Garman, 1917. *Nomen nudum.*
Geochelone nigra duncanensis, Pritchard, 1996.

***Geochelone nigra hoodensis* (Van Denburgh, 1907)**
(Maps 1, 4; Figs. 3, 11, 24)
Type locality: "Hood Island" (Española).
Testudo hoodensis Van Denburgh, 1907.
Testudo elephantopus hoodensis, Mertens and Wermuth, 1955.
Geochelone elephantopus hoodensis, Pritchard, 1967.
Chelonoidis hoodensis, Bour, 1980.
Geochelone hoodensis, Crumly, 1984.

***Geochelone nigra phantastica* (Van Denburgh, 1907)**
(Maps 1, 7; Fig. 20)
Type locality: "Narborough Island" (Fernandina) (probably extinct).
Testudo phantasticus Van Denburgh, 1907.
Testudo phantastica, Van Denburgh, 1914.
Testudo elephantopus phantastica, Mertens and Wermuth, 1955.
Geochelone elephantopus phantastica, Pritchard, 1967.
Chelonoidis phantastica, Bour, 1980.
Geochelone phantastica, Crumly, 1984.

***Geochelone nigra porteri* (Rothschild, 1903)**
(Maps 1, 6; Figs. 4, 16, 23, 27, 42–43, prologue, back cover)
Type locality: "Indefatigable Island." Southern Indefatigable Island (Santa Cruz). Name requires validation by the ICZN by suppression of the possible senior synonym *Testudo nigrita* Duméril and Bibron, 1835, whose holotype is of uncertain provenance.
Testudo porteri Rothschild, 1903.
Geochelone elephantopus porteri, Pritchard, 1967.
Geochelone porteri, Fritts, 1983.

***Geochelone nigra vicina* (Günther, 1875)**
(Maps 1, 10, 11; Figs. 5, 9, 24, 32, 48–58, prologue)
Type locality: unstated; given as "Iguana Cove, Süden der Insel Albemarle" by Wermuth and Mertens (1977). Southern and middle Albemarle Island (Isabela), from Iguana Cove, Cerro Azul, to Sierra Negra, Volcan Alcedo, and to Tagus Cove, Volcan Darwin.
Testudo vicina Günther, 1875.
Testudo güntheri Baur, 1889.
Type locality: unstated, given as "Villamiel, Südwesten der Insel Albemarle" by Wermuth and Mertens, 1977.
Testudo macrophyes Garman, 1917.
Type locality: "Santa Isabela Island (Albemarle) near Tagus Cove."
Testudo vandenburghi De Sola, 1930. *Nomen nudum.*
Type locality: "Forty miles from Villamil...at the coast on the southern border of Perry Isthmus [Albemarle Island]." (This population now extinct).
Geochelone elephantopus guntheri, Pritchard, 1971a.
Geochelone elephantopus guentheri, Pritchard, 1971b.
Chelonoidis elephantopus, Bour, 1980.
Chelonoidis guentheri, Bour, 1980.
Geochelone vicina, Crumly, 1984.
Geochelone vandenburghi, Crumly, 1984.

There are four possibly distinct subspecies for which no names are yet available:

***Geochelone nigra* ssp.**
(Maps 1, 5; Fig. 18)

Northeastern Chatham Island (San Cristóbal). Distinct from extinct southwestern and central Chatham Island populations.

***Geochelone nigra* ssp.**

(Maps 1, 6; Figs. 28–29)

Northwestern Indefatigable Island (Santa Cruz). Distinct from other Indefatigable populations, but may be an imported population, possibly from Duncan.

***Geochelone nigra* ssp.**

(Map 1)

Barrington Island (Santa Fé). Extinct, possibly introduced.

***Geochelone nigra* ssp.**

(Map 1)

Jervis Island (Rábida). Extinct, possibly introduced.

Efforts should be made to secure type material for the first two of these forms, at least. Eibl-Eibesfeldt (1959) reported that he had found skeletal material in northeastern Chatham in 1957, and that this has been deposited in the Senckenburg Museum (Frankfurt-am-Main). Perhaps this material would be adequate. Taking living tortoises from the Galápagos is now prohibited, yet voucher material — at least a holotype, such as a salvaged shell from a natural mortality — is essential if a description of a new form is to have legitimacy. Moreover, the need is not purely academic. Conservation investment tends to concentrate on named taxa, and the northwestern Indefatigable tortoises, although among the most distinctive in the archipelago, have received no conservation attention, despite their rarity, presumably in part because the population has not been nomenclaturally recognized as a subspecies.

SURVIVAL STATUS

The different populations of Galápagos tortoises have all come under varying degrees of threat to their survival during the last few centuries. Some populations have become totally extinct (e.g., Charles and southwestern Chatham), some are nearly or probably already extinct (e.g., Abingdon and Narborough), some have possibly been saved from extinction only by heroic conservation efforts (e.g., Hood and Duncan), and a few retain relatively robust populations (e.g., southwestern Indefatigable and Volcan Alcedo on Albemarle). All populations, however, continue to be at risk and extremely sensitive to the vagaries of human exploitation, feral and native animal depredation, and various environmental catastrophies. A constant vigil is required to monitor population levels and their changes through time if conservation management techniques are to be effective. The following section presents an historical review of the survival status of each of the separate populations of Galápagos tortoises, from the earliest recorded information up to as near the present as the published literature allows. This knowledge of former population patterns is critical to any understanding and proper interpretation of current dynamics as they relate to conservation and management issues. Ongoing census work by the Charles Darwin Research Station continues to monitor and document the status of these populations.

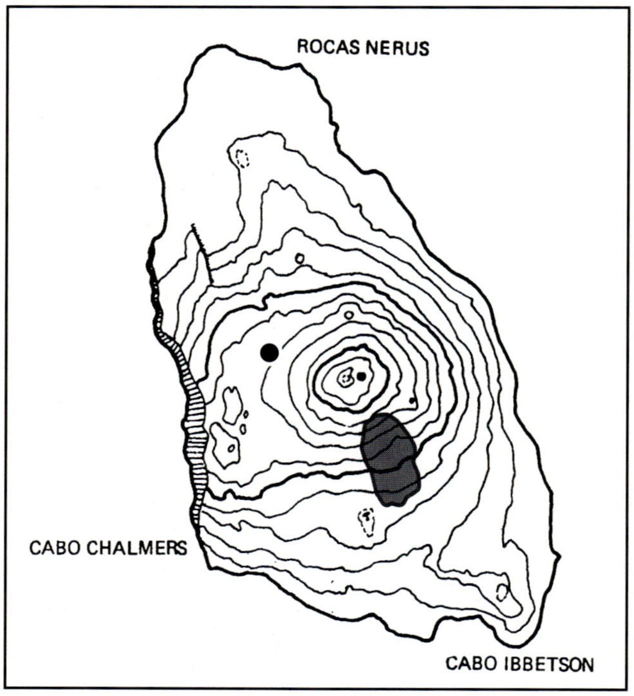

Map 2. Abingdon Island (Pinta). Shaded area is where tortoise skeletons have been found in ravines. The last living tortoise (*Lonesome George*) was found near the dot.

Historical Review of Galápagos Tortoise Populations

Detailed accounts of the status of the various populations of Galápagos tortoises include those of Groombridge (1982) and de Vries (1984). Most of the data quoted, however, were already several years old, and the most recent account based upon extensive first-hand information is that of MacFarland et al. (1974a). In 1989, a plan to recensus all of the tortoise populations was initiated by the Charles Darwin Research Station and the Galápagos National Park Service (L. Cayot, *pers. comm.*). As of 1995, censuses have been carried out for 10 of the populations and the remaining 5 are planned within the next two years. These studies will help to determine whether the intensive conservation efforts of the last 25 years are reflected in increased populations, and will provide valuable information on growth rates, survivorship, and overall population dynamics, in that useful numbers of individuals of most of the surviving populations were measured and marked (by shell-notching) in the 1960s and 70s.

Abingdon (Pinta). — This population, heavily exploited by whalers and other mariners, underwent an overall collapse in the mid-nineteenth century (Townsend, 1925b; Pritchard, 1977). The last recorded whaling crews to obtain large hauls of tortoises from Abingdon were those of the *Abigail* (142 tortoises in 1837), the *Hector* (67 in 1843), and

Figure 33. Habitat in the southern uplands of Abingdon Island, 1972.

Figure 34. *Lonesome George*, the last surviving tortoise from Abingdon Island, photographed just before removal from Abingdon in 1972.

Figure 35. Bony carapace of adult female Abingdon Island tortoise slaughtered by man. Note cut edges of shell and absence of plastron. This may have been the last female of this distinctive population. Carapace presented to the Charles Darwin Research Station by the author. April 1972.

two other ships that together collected 33 in 1848. Since then, no haul of more than 6 was recorded. In 1901, Rollo Beck collected 2 for the Rothschild collection (Nos. 1 and 27, an old male and an allegedly mature female, although the latter may have been an immature male). In 1905–06, the *Academy* party (also including Beck) found three live specimens and a shell, all males.

The status of the population between 1905–06 and the late 1950s is unclear. In 1957 M. Castro reported to Eibl-Eibesfeldt (1959) that tortoises were still to be found "far up the slope of the main volcano," and Castro and Eibl-Eibesfeldt found an old trail that seemed to be a tortoise trail at a relatively low altitude in 1957. However, in 1964 Castro and Cavagnaro were only able to find 28 skeletons of tortoises, all wedged or trapped in ravines and crevices on Abingdon, with no signs of living specimens (Snow, 1964). Moreover, because of the abundance of fissures and holes in the area, Snow added that "there must have been very many more that were not found." While this accumulation of skeletons may have represented normal, casual mortality over many decades (tortoise bones and shells are still found in caves on Charles Island, 150 years after the demise of the population), it is significant that goats had been introduced to the islands during the preceding seven years, and had multiplied extraordinarily. The implication is thus that the tortoises, under unaccustomed grazing pressure, had been forced into abnormal habitat (Fig. 33), and had died in accidents.

The population was deemed extinct, but a live adult male was seen by J. Vagvolgyi in 1971 (Pritchard, 1977)(Fig. 34), and in 1972 this individual was relocated and transferred to the Darwin Research Station for safekeeping, where it still resides, now known as *Lonesome George*. The circumstances of its rediscovery in 1972 are described by Cruz (1994).

Efforts to locate other individuals have proved fruitless. The goats on Abingdon have progressively been eliminated by shooting over about a decade (about 26,000 were shot in 1972, the first year), and the vegetation is now substantially recovered (Whelan and Hamann, 1989). I was able to find a carapace of an adult female on Abingdon in 1972, the plastron of which had clearly been severed by human action (Fig. 35). This was deposited in the Charles Darwin Research Station collection, and may be the only female specimen in any collection. A skull, cervical vertebra, and shell fragments of a large male (PCHP 50) were also found, again with the characteristic signs of having been slaughtered with a machete, and more recently D. Green

Figure 36. California Academy of Sciences party collecting tortoises on Duncan Island, 1906. Photo by R.H. Beck, reprinted from Slevin (1931). The original caption read: "We had a luxurious camp on Duncan Island. The pet hawk, standing on the dead tree, was fed turtle meat every day and became quite tame."

(*pers. comm.*) and companions found a very large male shell, with some scutes still attached; this too is now at the CDRS.

In March 1981 (i.e., nine years after the removal of *Lonesome George*, the supposed last surviving Abingdon tortoise), Ole Hamann and Linda Cayot found a single old dried tortoise scat on Abingdon Island (Anonymous, 1982). In that the scat was intact and lying on top of dried grasses (L. Cayot, *pers. comm.*), and furthermore noting that several El Niño (high rainfall) seasons had intervened between 1972 and 1981, the possibly exists that *Lonesome George* is not the last.

Townsend (1925b) includes a photograph of a live adult Abingdon tortoise that was "among the first... brought to the New York Zoological Park," but gave no details of collection. Baur (1889) alludes to an old male from Abingdon that had been collected by the *Albatross* expedition, and that ended up, "nearly complete," in the USNM collection. Perhaps this was the same one.

Duncan (Pinzón). — This population was reasonably abundant in the nineteenth century, having survived the occasional onslaught of victualling whalers, but suffered heavy collection for scientific purposes (e.g., 8 collected by Baur in 1891, 29 by the Webster-Harris expedition in 1897, 86 by the *Academy* expedition in 1905–06; Fig. 36). It was feared virtually extinct by the 1920s (Beebe, 1924), and indeed Townsend, combing the slopes of Duncan in 1928 with the assistance of eight men, did not find a single tortoise, and he suggested that they were already extinct. But a small population of aging individuals survived (Figs. 1, 2, 37, 38), although poaching also continued — in 1957, Eibl-Eibesfeldt (1959) collected the shell of an adult male that had been killed about a year earlier.

Successful reproduction of the Duncan tortoise apparently ceased in the 1890s when feral black rats reached the island (first recorded in 1891 according to Patton et al., 1975) and preyed upon the hatchling tortoises with nearly 100% efficiency. The resulting absence of young was already noticed by Beck (1903), and was reflected in the abrupt truncation of the lower-end size classes of the 1905–06 tortoises, the smallest of which had carapace lengths of about 47 cm. Indeed, in the years immediately following their introduction, there is evidence that the rats, even in the presence of numerous short-eared owls, were particularly voracious and would even attack large tortoises if the latter were confined or compromised in some way. Thus, when Walter Rothschild enlisted the services of Webster and Harris in 1898 to collect tortoises for him on Duncan, Harris records in his diary (*in* M. Rothschild, 1983) the following description of the party's excursion ashore to pick up eight tortoises that had been located an immobilized with heavy stones two days earlier: "On getting to the crater, we found one big tortoise dead; one of the big rocks that we weighted it with had shifted and fallen on its neck, and shut its wind off. Rats had gnawed out one of its eyes and had also gnawed a piece out of one hind foot of the living smaller specimen."

Since 1965, eggs from natural nests have been taken to the Darwin Station for hatching and captive head-starting,

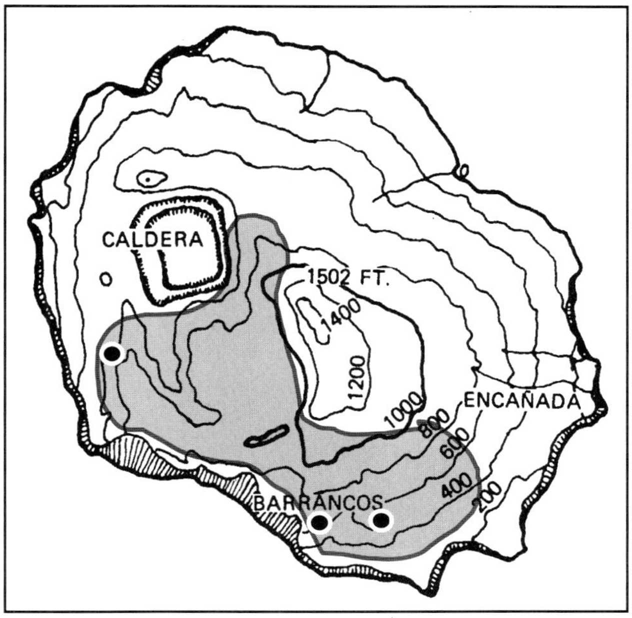

Map 3. Duncan Island (Pinzón). Shaded area is overall distribution of native and repatriated tortoises. Dots are nesting areas.

and the wild population of 150–200 old adults (MacFarland and Reeder, 1975) had been supplemented by 226 young by 1985 (Tierney, 1985) and 268 by December 1990 (Caporaso, 1991), although in a few cases the young tortoises, released too early, have been killed by rats (de Roy Moore, 1979; de Vries, 1984).

By 1990, the native adult population was down to 80–100 (Morillo and Cayot, 1990), with a few more dying each year (Figs. 15, 39), at least 8 during the drought of 1984–85. Indeed, of the 64 females marked by shell-notching from 1963–69, only about sixteen have been recorded since 1977, although this finding may stem, in part, from both healing and abrasion of the peripheral shell notches occurring over the years, so that the coded notch combinations can only be identified in less than 50% of individuals (L. Cayot, *pers. comm.*).

Márquez et al. (1992) report the numbers of Duncan tortoises hatched at CDRS from 1965–66 to 1986–87, and Caporaso (1991) gives the numbers repatriated for each year from 1970 to 1990. Repatriated tortoises are shown in Figs. 12, 40, and 41, and total numbers repatriated are given in Table 1.

In the early 1970s around 50 nests were found each nesting season, but by the end of the decade the number was closer to 20 per year, and from 1980–86 fewer than 10 were found annually in the traditional nesting areas (which incidentally suffered much erosion by rain between December 1982 and July 1983). Nevertheless, this is not necessarily a sign of actual (adult) population collapse, in that during several of these nesting seasons, the park wardens involved in the surveys were new and relatively untrained, and also several extreme drought years unquestionably depressed the nesting productivity. Indeed, it was extremely encouraging when more experienced personnel were able to locate no fewer than 45 nests in September 1987 (Fritts, 1988), the

Figure 41. Healthy, large subadult repatriated tortoise on Duncan Island. Note the contrast in anterior shell profile and elevation of the carapace with the native adult male in Fig. 38. September 1982.

majority in the more western nesting zone, and by the end of the 1987–88 season the total had risen to 96 nests (Morillo, 1992; Cayot et al., 1993), although extreme drought the following year reduced the number to just 11 nests found. In 1989–90, 29 nests were found. In that tortoises are not normally seen in the course of nesting, it is uncertain what proportion of the nests in recent years have been made by native tortoises, as opposed to repatriated animals, but L. Cayot (*pers. comm.*) is of the opinion that the majority are still made by natives.

The growth rates of the repatriated animals have been very variable. In some years, no growth at all occurred for ten months following release, whereas other year-classes showed uninterrupted growth (Metzger and Marlow, 1986). The oldest repatriated tortoises, from 1965–66, have grown at excellent rates, even surpassing native adults in size within two decades (Rhodin et al., 1983; Pritchard, 1985), and Caporaso (1991) reported signs of breeding and preliminary nest building. The repatriated tortoises have shown an apparent sex ratio of 0.63:1 males to females, contrasting with the wild sex ratio of 2:1 (Metzger and Marlow, 1986), although Morillo (1992) found the native tortoise sex ratio to favor females.

A conservation project can only be considered complete when the causes of the crisis or the decline have been addressed directly. Thus, the head-starting program for the young Duncan tortoises was essential to prevent the native population from disappearing through the period of almost 100% recruitment failure. But the Charles Darwin Research Station and the Galápagos National Park Service did not miss the opportunity, following the extreme drought of 1988, to attempt to eliminate the feral black rats (*Rattus*

(Figures on facing page)

Figure 37 (upper left). Adult male native Duncan Island tortoise, sheltering from the sun. September 1982. Photo by A.G.J. Rhodin.

Figure 38 (upper right). Very old adult male tortoise on Duncan Island. This individual, known in the islands as *Onan*, has since died. Note the pitted surface of the carapace, indicative of fenestration in the bony shell, and the growth of lichens in concave parts of the carapace. This same individual is also depicted in the painting on the front cover. August 1986.

Figure 39 (lower left). Carapace of a female tortoise on southwestern Duncan Island. Every year sees the natural death of a few more of the native stock of very old tortoises on Duncan. September 1982. Photo by A.G.J. Rhodin.

Figure 40 (lower right). A tall *Opuntia galapageia* var. *macrocarpa* cactus in the southern highlands of Duncan Island provides minimal but adequate shade for a young, repatriated tortoise. September 1982.

rattus), whose depredations upon the hatchling tortoises constituted the primary threat to the population. The time seemed especially opportune in that the dry conditions had not only reduced the rat population to the lowest level in many years, but also the presumed extreme hunger and thirst of the survivors might predispose them to be attracted to appropriate bait.

The campaign was described by Cayot et al. (1993). It took place initially on a localized scale, in the critically important western tortoise nesting zone. Subsequently it was expanded into five work zones that covered the whole island. Total personnel numbered 47, and bait stations were established at 50 m quadrants over the entire island. Poison used included the commercial rodenticide Klerat, as well as the anticoagulant Racumin mixed with rice.

The campaign clearly reduced the rats to a very low level indeed, and in the course of 1989 the only evidence of surviving rats was found at a single site on the highest part of the island. Nonetheless, success cannot yet be claimed, rats having extraordinary powers of reproduction once favorable conditions return. But the virtual elimination of the rats promoted healthy increases in populations of small native vertebrates, including lava lizards, marine iguanas, Galápagos doves, and other small bird species. Significantly, however, a number of dead hatchling tortoises were found, 12 in March 1988 and 9 in late 1989, with small holes in the carapace suggestive of predation by native hawks (*Buteo galapagoensis*) rather than by introduced rats (Cayot et al., 1993). It always presents a frustration to conservationists when one rare species attacks another, and even more so when it is possible that the predation upon the rare native species may have been a result of the virtual elimination of the previous prey — the harmful feral species.

Table 1. Numbers of repatriated Duncan Island tortoises.

Season of hatching	Number repatriated	Year of repatriation
1965–66	20	1970
1965–66	9	1971
1966–67	20	1971
1967–68	23	1971
1968–69	10	1972
1969–70	29	1973
1970–71	11	1974
1971–72	17	1976
1972–73	18	1976
1973–74	14	1977
1974–75	11	1979
1975–76	10	1980
1976–77	3	1981
1977–78	4	1983
1978–79	1	1983
1979–80	26	1984
1980–81	13	1985
1981–82	7	1985
1982–83	–	–
1983–84	11	1987
1984–85	11	1988
1985–86	–	–
1986–87	23	1991?

Following the intensive field work on Duncan in the late 1980s, Cayot et al. (1993) reported that 258 tortoises had been observed in the course of the census, out of an estimated total of 292 animals, the total presumably including both native and head-started or repatriated animals, but not hatchlings or post-hatchlings. Of the 100 native tortoises that had been marked in the 1960s, 73 were found, although in many cases the original shell-notch combinations were no longer readable.

Hood (Española). — The whalers reduced these tortoises drastically. Hood is a small, low-lying island, and is the first of the islands to be encountered by mariners arriving from around Cape Horn, and the tortoises, being among the smallest in the archipelago, were easy to carry. Townsend (1925b) found records of at least 1698 tortoises removed from Hood from 1831 to 1868, the large hauls being towards the beginning of this period. The *Isabella*, for example, collected 335 in five days in 1831, and the *Loper* obtained 237 in four days in 1834. But by 1842 the news was bad. The Master of the *George Washington* reported that "Terrapin [were] very scars" on Hood in 1842, and in 1847 the *Persia* "got but few turpin," although the *Charles Frederick* was able to collect 67 from Hood in that year. But thereafter the take was never more than 7 on a given visit.

By 1905–06 the whalers were long gone, but instead feral goats, present in thousands, had denuded the landscape, virtually eliminating food or protection for young tortoises and competing dangerously with the surviving adults for food. The *Academy* expedition in those years only obtained 3 live tortoises on Hood despite protracted searches, and the laconic conclusion was "evidently they have been well cleaned out."

Nonetheless, some survived. Banning (1933) reported that the Pinchot expedition of 1929 had recalled the "extinct" Hood tortoise to the status of "rarity," although the two specimens encountered were both removed, running a serious risk of provoking a return to the former status (Thornton,

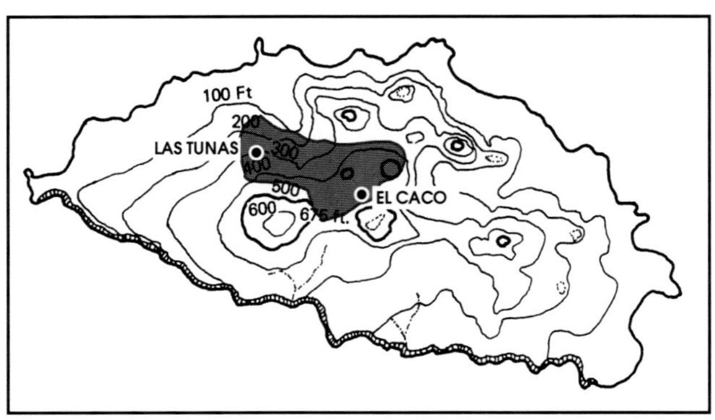

Map 4. Hood Island (Española). Shaded area is overall distribution of repatriated tortoises.

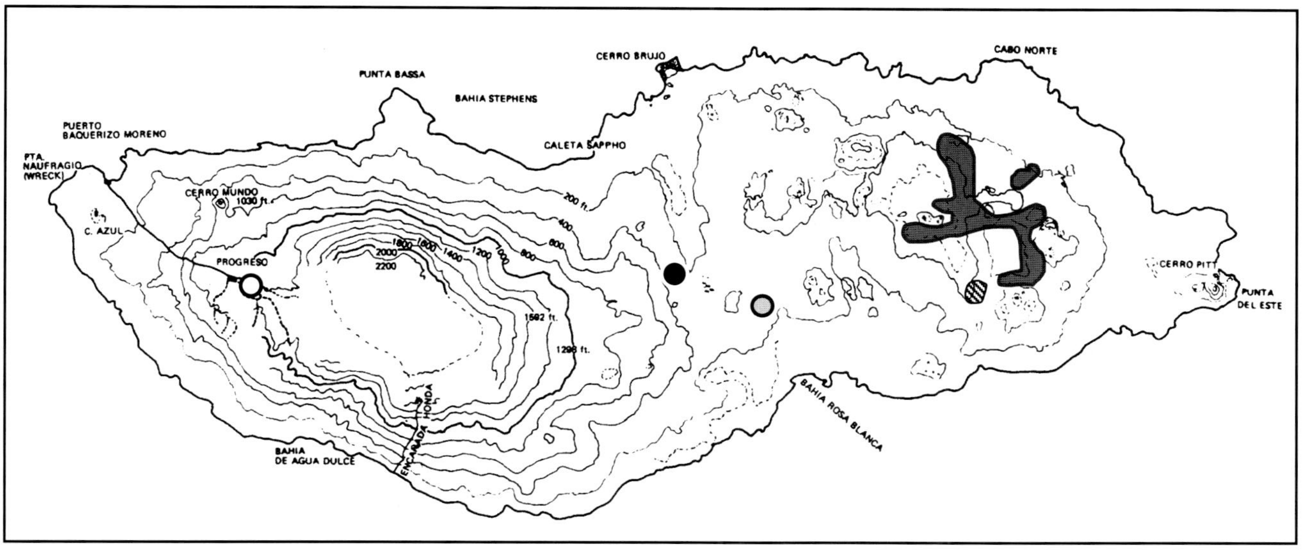

Map 5. Chatham Island (San Cristóbal). *Northeastern population*: Shaded areas represent low density tortoise distribution, striped area is a region of high tortoise density. *Southwestern and central population*: The black dot is the approximate location of the last live tortoise collected in 1906, the lightly shaded open circle is a cave where 17 skeletons were found in 1906, the open circle marks the approximate location where the last surviving specimen may have been killed in 1933.

1971). There was some level of US military presence on Hood during WWII, but records are not available as to whether tortoises were encountered. But Eibl-Eibesfeldt (1959) stated that a shell with scutes intact, that he found in 1957, was from a tortoise killed in 1942.

The tortoises located on the island during the 1950s and 60s were so few and dispersed that it is assumed that reproduction did not occur (Corley Smith, 1977), although it is also possible that hatchlings were produced but that they starved, baked, or failed to survive the onslaughts of hawks and other predators in a goat-cleared landscape. The first animal located in recent decades was an apparent male, 56 cm in carapace length, found feeding on a fallen *Opuntia* cactus in competition with fifteen goats by M. Castro on 31 August 1963 (Snow, 1964). On 24 November 1963, another individual, 59 cm in length, was located, and the first specimen was re-sighted in late July 1964. Snow estimated that the total population could not be above about 20 individuals, and he was almost certainly correct. The decision was made to transfer all live tortoises that could be found to a captive breeding center on Indefatigable Island, and ultimately a group of 3 males (one a long-term captive at the San Diego Zoo) and 12 females were accumulated at the Darwin Station. Reproduction has been excellent. By 1982, 113 young had been returned to Hood Island (Reynolds, 1982), and the feral goats had been eliminated by shooting. By 1986, 184 had been repatriated, of which only 7 had been found dead (Anonymous, 1986).

Figures given by Márquez et al. (1992) indicate that, between 1980–81 and 1986–87, the captive colony laid between 24 and 32 nests annually (138–204 eggs). Percentage hatching was low (9.3–33%) but a total of 206 individuals from the year-classes 1970–71 to 1984–85 had been released on Hood Island as of 1992, with 53 animals from 1984–85 and 1985–86 still in captivity. The earlier released animals have reached adult size (pers. obs.), and a recent report (Caporaso, 1991) indicated that 328 Hood Island tortoises had been repatriated as of January 1991.

Updated figures are given by Cayot and Morillo (in press). They reported that releases had occurred at two sites (El Caco and Las Tunas) in the interior of Hood Island. By the end of 1993, 575 tortoises had been released, with 288 more still held in captivity at the Breeding and Rearing Center at the Charles Darwin Research Station. The survival rate of the very young tortoises (released at 12–18 months of age) was poor — Piños (1987) indicated a mortality rate of 56% within the first year. One tortoise was apparently killed by a native hawk the day following its release. But between 1988 and 1992, the minimum survivorship was 55%, and a total of 202 different tortoises was observed on the island (average of 89 individuals observed per trip). Four nest attempts were found in 1990 (Márquez et al., 1991), all at El Caco, two carcasses of hatchlings killed by hawks were found in late 1990, and a live hatchling was found in June 1991. Sixteen nests were found in November 1991, and 12 nests (and two live hatchlings) in 1993.

Chatham (San Cristóbal). — The tortoises of southwestern and central Chatham Island were especially heavily collected by whalers in the mid-nineteenth century — Townsend (1925b) found that, of 13,013 tortoises reported to have been collected by whalers in the Galápagos from 1831 to 1868, 4798 were from Chatham. A minute remnant of this population survived to 1905–06 (Van Denburgh, 1914), but it is now extinct. Possibly the last was killed by settlers near Progreso in 1933 (Banning, 1933).

Nevertheless, several hundred tortoises still live in the arid northeastern part of Chatham Island, although they are apparently a different subspecies. Feral dogs have menaced this population in the past, but may now have been eliminated. In addition, captive hatching and head-starting programs have resulted in the release of 13 animals from the

1972–73 year-class, 6 from 1973–74, 23 from 1974–75, 5 from 1975–76, and 8 from 1978–79 (Márquez et al., 1992). Gordillo and Kendrick (1989) pronounced that the tortoise population of northeastern Chatham (Fig. 18) was one of the most secure in the archipelago, following a decade of diligent research and management activities by National Park personnel.

Indefatigable (Santa Cruz). — This large, central island, today inhabited by several thousand people and the headquarters of the Charles Darwin Research Station and the Galápagos National Park Service, was settled on a temporary basis before 1846, when huts occupied by gatherers of archil (a lichen used in the dye industry) were built at Whale Bay on the west coast, with a trail leading inland to small citrus groves at Santa Rosa and Salasaca (Perry, 1972). These huts were finally abandoned (although still to be seen at the turn of the century), and Indefatigable did not receive permanent inhabitants until the Norwegian colonizing expedition of 1926. Tortoises from Indefatigable featured from time to time in the logbooks of American whalers — Townsend (1925b) reported that 44 were collected by the *Pacific* in 1833, 140 by the *Abigail* in 1834, 12 by the *Benezet* in 1834, 2 plus "many" by the *Pioneer* in 1836, 45 by the *Alfred Tyler* in 1845, and 36 by the *Roman* in 1848. But in general the collecting pressure was slight — of 13,013 tortoises reported collected in the Galápagos between 1831 and 1868, only 279 were from Indefatigable.

Cookson (1876) claimed that tortoises were so rare on Indefatigable in 1875 that they were no longer hunted, and Rothschild (1903) incorrectly claimed that, since the 1812 visit of David Porter to the Galápagos, "no one has found tortoises on Indefatigable, and it was supposed they had been exterminated." But he goes on to report that Beck, under contract to Rothschild, had found a young tortoise near the shore of Indefatigable in 1901, and in 1902, after a long and wearisome hunt, Beck collected 7 more, ranging up to very large size. The *Academy* expedition in 1905–06 collected 23, and concluded that the Indefatigable tortoise was still fairly common (Fig. 42).

In the years following permanent settlement, slaughter of tortoises on Indefatigable was widespread. A 1930 expedition to the highlands of Indefatigable, described by Townsend (1930) and Roosevelt (1930), found just two of the 35 Norwegian colonists who had arrived four years before still in residence on the island, and they and some Ecuadorian settlers regularly hunted the tortoises. Townsend's party found a single animal weighing an estimated 250–300 pounds. Local people brought in 8 smaller ones, 40–80 pounds in weight, and these were taken to the

Figure 42. Until trails were made into the tortoise area on Indefatigable Island, access was very difficult, and few tortoises were removed. Photo by R.H. Beck, reprinted from Slevin (1931). The original caption was: "Packing tortoises through the jungle of Indefatigable is no child's play."

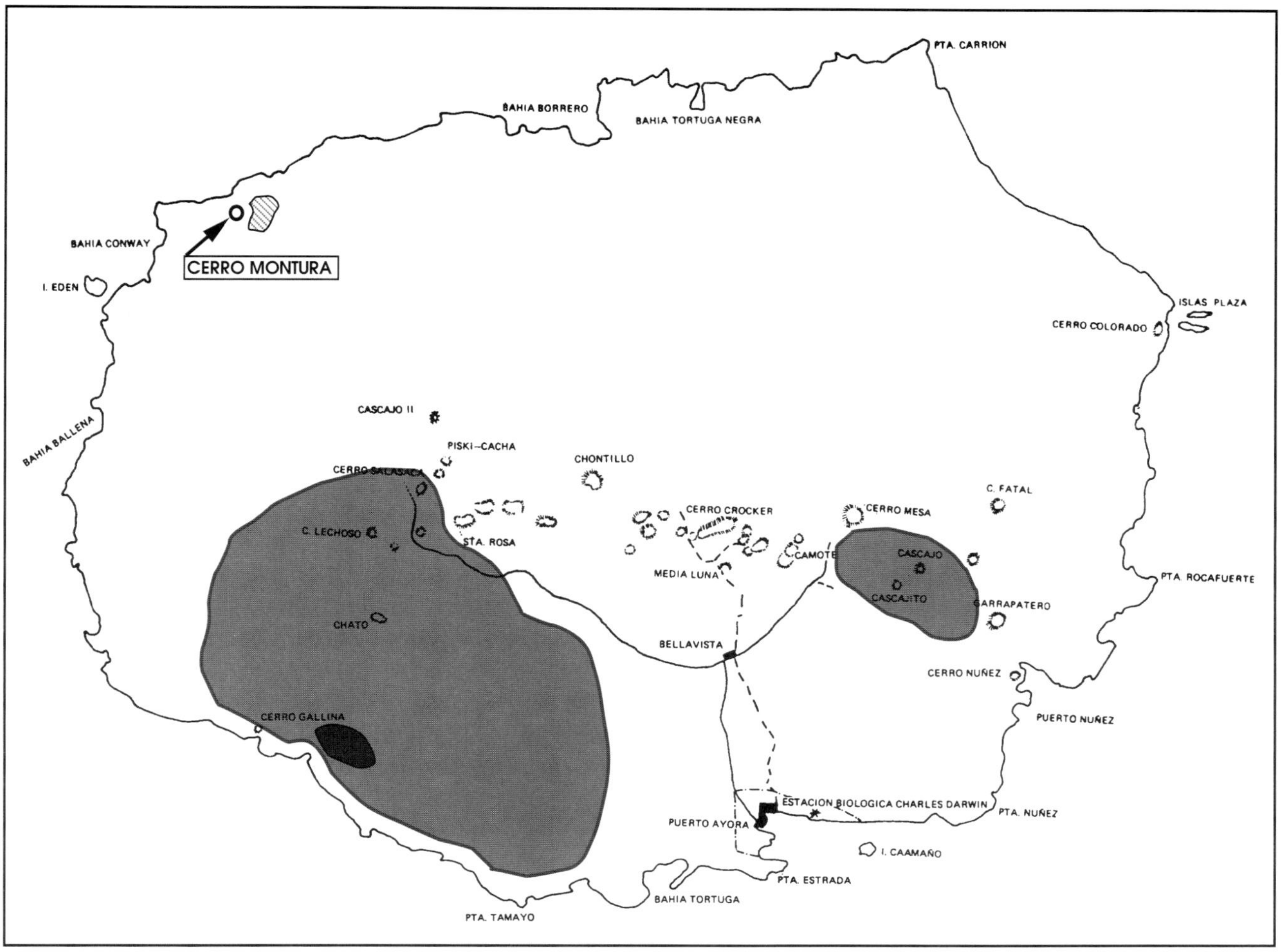

Map 6. Indefatigable Island (Santa Cruz). Lighter shaded areas are the distributions of the dome-shelled southern tortoise populations, the darker shaded area is a nesting zone. The striped area at Cerro Montura is the distribution of the saddlebacked northwestern population.

New York Zoological Park. The party also learned that natives of Chatham Island had come to Indefatigable eight months earlier and killed a number of tortoises. Townsend wrote: "While in the tortoise country we measured the bleached shells of large tortoises killed years before. Some of them were nearly five feet in length of top shell as measured over the curve. They must have weighed considerably over 500 pounds... The Norwegian settlement on this island, recently abandoned, must have killed many tortoises as we found numerous large skeletons during our brief stay in the tortoise country... Tortoise meat was still an important part of their food supply and tortoise oil served them as butter. They informed us that young tortoises were seldom found. A five-pound specimen represents the only small tortoise we obtained. The Galápagos hawk, which is more abundant on Indefatigable than elsewhere, was credited with destroying newly hatched tortoises. It is possible, since tortoises are no longer abundant, that hawks serve to hinder their increase on Indefatigable." Today the hawk is almost extinct on Indefatigable, a few individuals surviving in the extreme northwest.

Rambech (*pers. comm.* in MacFarland et al., 1974a) estimated that at least one to two thousand tortoises were killed on the island by oil hunters in the 1930s. By mid-century, Indefatigable had been settled continuously for several decades, and the reports of several visitors suggested that tortoise hunting was still a problem. Eibl-Eibesfeldt (1960), visiting the *galapaguera* in southwestern Indefatigable in 1957, reported evidence of intensive slaughter: "...what we found was a graveyard. The shells of slaughtered tortoises lay scattered everywhere. Most of these carcasses were old and half-rotted away, though some were quite fresh. So this area was still ravaged by present-day settlers, despite all the prohibitions." Eibl-Eibesfeldt (1959) interviewed settlers on Indefatigable, and was advised that, 25 years earlier, tortoises had been found frm the arid zone up to the transition zone, although they were less abundant on the northern slopes than the southern. Today they are confined to the southern slopes (apart from the disjunct colony in the extreme northwest). He also found bones and shell fragments in a cave about two miles north-west of Academy Bay, an area from which they have since disappeared. And he mentioned a few records from the northern and eastern parts of the island. He concluded that the species was "in rapid retreat."

Hendrickson (1965) illustrated the shells of slaughtered tortoises found on Indefatigable, but gave no details. MacFarland et al. (1974a) indicated that 15–25 tortoises

Figure 43. Adult male tortoises in a drying pond in the southeastern uplands of Indefatigable Island. July 1986.

were slaughtered annually on Indefatigable during 1965–68; 15 in 1969; 6 in 1970; and 2 in 1971. Thereafter there has been little killing of these tortoises, the result of law enforcement and educational efforts by the Charles Darwin Research Station and the Galápagos National Park, but various other stresses remain. In certain years, for example, there has been heavy predation upon the nests of Indefatigable tortoises by feral pigs (subsequently brought under control).

Black (1973) noted that he had encountered carapaces of 11 tortoises killed by man in the eastern section of Indefatigable Island, and he estimated that about 20 tortoises were slaughtered annually in the years up to 1971. Prior to the establishment of the National Park, a Tortoise Reserve had been declared in southwestern Indefatigable, in the area south of the village of Santa Rosa and east of Puerto Ayora, adjacent to the highland agricultural and pasture areas. The slaughtered animals were found both inside and outside the Reserve. Black described how the tortoises were brutally chopped open while still alive. Interestingly, however, the shells of reportedly slaughtered animals in the accompanying photograph were complete (i.e., with plastron still attached, and no obvious machete marks visible).

As early as 1960–61, tortoises on Indefatigable were being marked with painted numbers by a local worker answerable to the Director of the Charles Darwin Research Station, and about 1000 animals were censused in this way. But paint was unlikely to last for more than a few months, and subsequently painted numbers were used only for short-term studies where the behavior of specific tortoises needed to be observed from a distance.

Hendrickson (1965) reported that 700 Indefatigable tortoises had been marked by carapace-notching, and that the population was estimated to number between 1500 and 2000 individuals. De Vries (1968) indicated that 371 males, 367 females, and 254 juveniles (992 total) had been marked, and he estimated the actual population to number 2000–3000 individuals. Groombridge (1982) estimated 3000–4000, as did Cayot (1985), who indicated that more than 2000 had been marked. The vast majority of the population is concentrated in the southwestern quadrant of Indefatigable, but there is an outlying population, now separated from the main population by a settled agricultural zone, to the east, as well as an enigmatic saddlebacked population, very different in external morphology, in the far northwest.

The tortoises are strongly migratory. Generally speaking, large tortoises move to the highlands, near the village of Santa Rosa, at dry times of the year, and some remain there year-round (Fig. 43). Migrations to the lowland nesting zones take place generally in January to May (Rodhouse et al., 1975), and very young tortoises are thus usually found at low altitudes. However, extraordinary climatic events can precipitate extraordinary movements, and Cayot (1985) documented a dramatic exodus of tortoises from the highlands to the lowlands during the El Niño event of 1982–83, when torrential rivers flowed in normally arid country, and when even the lowland nesting areas were so wet that many nests were abandoned prior to completion.

Snow (1964) drew attention to the danger of extinction of the population of saddlebacked tortoises in the area of Saddleback Hill (Cerro Montura) in northwestern Indefatigable. Intensive searches since the discovery of this population in 1962 had revealed only 7 individuals (an adult male and six juveniles, less than five years old), plus a dead very young specimen. The population survived in the 1970s and 1980s — I encountered a young tortoise in 1972 (Fig. 29), and reasonably fresh droppings in 1982 (Fig. 30), and R. Reynolds (*pers. comm.*) reports having encountered two individuals of medium size. Study and conservation of this population is now a priority for the Galápagos National Park Service.

The origins and taxonomic status of this population remain obscure. One tantalizing scrap of data is provided by a photograph, published by Fritts and Fritts (1982), of an unidentified member of the *Academy* expedition of 1905–06 (possibly Slevin) carrying what appears to be a small, smooth-shelled saddlebacked tortoise on his shoulders. The caption suggests that the photograph was taken on Duncan or Hood, presumably because both of these islands had small saddlebacked tortoises. However, in the background of the photo rises what appears to be the characteristic profile of Saddleback Hill, the chief landmark in the tortoise habitat of northwestern Indefatigable, and the presence of what appears to be a white sandy beach in the right foreground is compatible with this conclusion.

Many speculations are possible, one being that this tortoise (with others?) was taken by the *Academy* party from nearby Duncan Island, and released near Saddleback Hill, possibly constituting the maternal parent of the tortoises found there today. The photographs (Snow, 1964) (Fig. 28) of a large male (*King Gustav*) seen by Gusch Angermeyer in 1962 suggest a larger form than is found on Duncan today, but this could be a result of better nutrition on this larger and higher island, where even lowland tortoise populations have access to humid uplands during drought periods.

Narborough (Fernandina). — The surviving records of the whalers and buccaneers are essentially free of records of tortoises on Narborough. The single early mention was that of Porter (1822), who reported tortoises in greater or lesser abundance on all the larger islands of the group that he visited, namely "Hood's, Marlborough, James, Charles, and Porter's." He also found tortoise bones on Chatham, but was on Albemarle for too short a time to find tortoises, and he did not go to Bindloe, Abingdon, Downe, or Barrington. In that Porter's Island is a synonym for Indefatigable, and Downe an apparent synonym for Duncan or Tower (Woram, 1989), it appears that all the larger islands are accounted for, and Marlborough must be a misspelling of Narborough, the

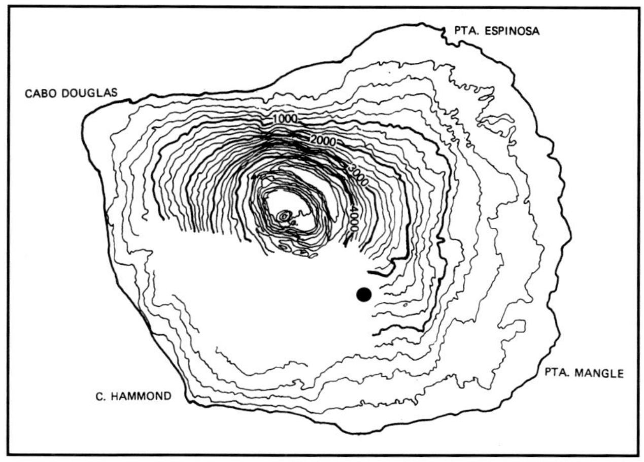

Map 7. Narborough Island (Fernandina). The dot represents the approximate location of the single tortoise collected by Rollo Beck in 1906.

name given to the island by Ambrose Cowley in 1685. The island has never been inhabited, and remains free of feral mammals. Yet the single adult male tortoise collected by Rollo Beck in 1905 (Fig. 20) remains the only tortoise ever seen on this island. Eibl-Eibesfelt (1959), in a 1957 walk clear across the island from three miles west of Cabo Douglas (he must have meant east — Cabo Douglas is the westernmost point of Narborough) up to the caldera and thence down to Punta Espinoza, found no trace of tortoises. In 1964 Hendrickson (1965), with the advantage of a helicopter for exploration of remote areas, located a large tortoise dropping and an *Opuntia* cactus with bite marks seemingly made by a tortoise. Possibly one or two individuals survive. The reason for the extreme rarity or extinction of this form cannot be ascribed to human activities. Conceivably the sporadic major eruptive activity of the Narborough volcano has destroyed nesting areas or separated them from feeding areas. Many of the large numbers of land iguanas (*Conolophus*) on this island today nest in ash deposits at the bottom of the vast caldera (Werner, 1982), a habitat inaccessible to tortoises.

Charles (Santa María or Floreana). — There is no evidence that this population survived beyond the 1840s. Heller (1903) estimated that extinction occurred in 1840; Broom (1929) estimated 1850. Steadman (1986) reviewed all available information, and thought that 1850 was a very reasonable estimate, following a decade of extreme rarity. By then the island had been sporadically inhabited for some time — a wild Irishman lived there alone at the beginning of the century, and a highland settlement was established by General Villamil in 1829, although troubles began when this was converted to a penal colony, and it was eventually abandoned. Later a new penal colony in the same area also failed.

Charles Darwin, on a visit to Charles Island in 1835, saw no live tortoises, only empty carapaces (Sulloway, 1984). Three years later, the frigate *La Venus* visited, but found no tortoises, and B. Seemann of HMS *Herald*, visiting in 1846, reported the tortoises to be extinct. Banning (1933) mentioned that a Charles Island settler (one of only eight on the

Figure 44. Mounted specimen allegedly representing the extinct Charles Island (Santa María or Floreana) tortoise in the British Museum.

island) claimed to have discovered a small tortoise that had since died, and this claim was upheld by neighbors. Yet, at that time, the settlers on Charles Island were a group of celebrated — or in some cases notorious — settlers who had arrived from Germany in 1929, who would surely have mentioned tortoises in their memoirs if they had found any (Strauch, 1936; Wittmer, 1960). Our knowledge of the morphology of the tortoises of Charles Island is derived from shells and other bones collected in caves near Post Office Bay, e.g., by Townsend (see Broom, 1929). In addition, Shurcliff (1930) reported the collection of several dozen tortoise shells for the Field Museum by himself and K.P. Schmidt in a cavern adjoining that visited by Townsend. And Steadman (1986) found abundant frag-

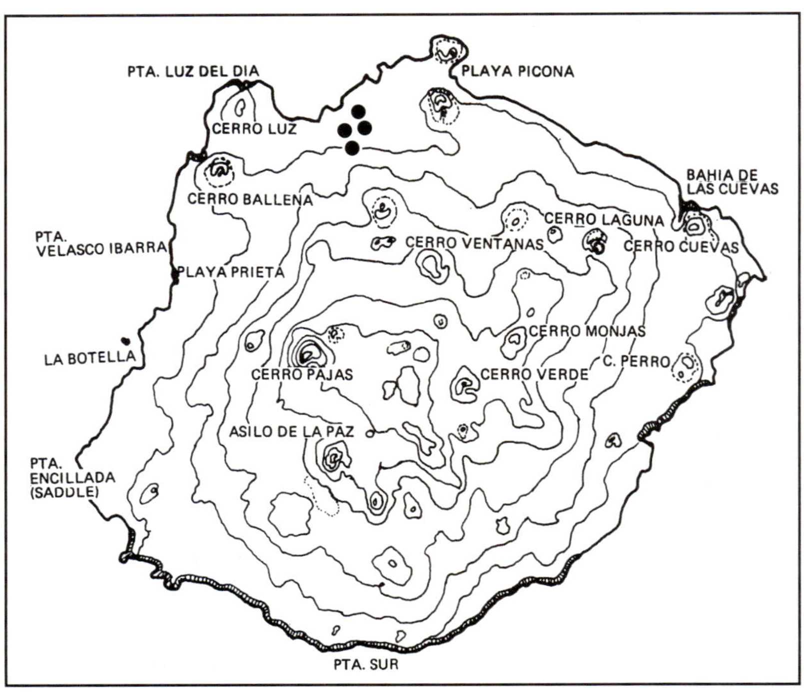

Map 8. Charles Island (Santa María or Floreana). The dots represent caves where subfossil tortoise remains have been found.

Figure 45. Bony carapace of the extinct Charles Island tortoise (adult female) collected in a cave on the island (Field Museum, Chicago).

ments in four cave systems: Finch Cave, Barn Owl Cave, Cueva de Post Office (Superior), and Cueva de Post Office (Inferior). He also reported small numbers of tortoise fragments in other, shallow caves in the Bahia de las Cuevas region. Apart from these subrecent cavern specimens and fragments, there is question — or absolute doubt — about all of the other specimens alleged to be from Charles Island (e.g., those illustrated by Garman, 1917) (Figs. 44–45).

The seemingly abrupt extinction of the Charles Island tortoise is curious. On islands of comparable size like Abingdon and Hood, analysis of the whalers' collecting records (compiled by Townsend, 1925b) indicates quite precisely the year in which these tortoise populations collapsed to a low level, but actual extinction was avoided by the extreme difficulty of locating the few, scattered survivors. Yet the records for Charles Island show substantial collections in the years immediately preceding apparent total extinction. In 1831 the *Magnolia* took 155 and the *Frances* took 179; the *Hector* collected 226+ in 1832; the *Octavia* 235 in 1833; the *Bengal* collected 100, the *Moss* 350, and the *Benezet* 120 in 1834; the *Barclay* took 50 and the *Benezet* 40 in 1835; and the *Eliza Adams* obtained 24 in 1837 (Townsend, 1925b; Steadman, 1986). Moreover, these records represent only a fraction of the visits by whalers to Charles Island during this period — there were 31 in just the ten-month period beginning October 1832 (Reynolds, 1835). Subsequent to this, the only records are a couple of unquantitative ones for the late 1840s — the *Congaree* logged "some Terrapin" from Charles in 1847, and the *Aurora* "got a few Turrapin" in 1848, and Townsend (1925b) concluded that these must have been tortoises imported from other islands for sale to mariners. The entry for the *Fabius* in 1858 ("boats ashore after turtle"), in its avoidance of some form of the usual vernacular "terrapin," probably referred to the East Pacific green turtles (*Chelonia agassizi*) that still nest on Charles Island.

Doubtless the total extinction of the Charles Island tortoise was hastened by the presence of the penal colony, the relatively small size of the island, the conversion of the highlands to farming and fruit plantations, and the influx of numerous feral mammals, including pigs and dogs. Pigs, cats, dogs, goats, burros, and cattle were all introduced around 1832 (Hoeck, 1984); the black rats and house mice may have arrived around the same time.

Jervis (Rábida). — As has been discussed above, only one live tortoise is known to have been collected on this island (Van Denburgh, 1914), and it may have been introduced. Although certain authors have associated the name *Testudo wallacei* (or subsequent combinations, including *Geochelone elephantopus wallacei*) with the alleged tortoise of Jervis Island, the arguments in favor of this are hardly acceptable. Nevertheless, the conclusion that Jervis never had native tortoises has to take into account the comments of Steadman and Zousmer (1988), mentioned above, frustrating in their lack of citation or any other basis for certainty, but nevertheless alluding without any expression of doubt to a former population of both tortoises and land iguanas on Jervis Island.

Barrington (Santa Fé). — The tortoises of this island are somewhat of a mystery. The island today has not had native tortoises in living memory, yet it is free of feral mammals, and such vulnerable species as the endemic land iguana and the rice rat have survived in good numbers. Townsend (1925b) found only two accounts of whalers having found tortoises on Barrington — 22 were taken by the *George and Susan* in 1839, and 1 by the *Henry H. Crapo* in 1853. Van Denburgh (1914) reported that no recent visitor to the archipelago had made note of tortoises on Barrington, although he then cited two informants, one of whom claimed to have taken tortoises off Barrington 16 years earlier (i.e., around 1890), and the other, a captain of an inter-island schooner, who claimed that tortoises had been found all over Barrington 30 years earlier (i.e., around 1876), some of which he had collected, but he doubted if any were left.

Possibly these observations were accurate, but if they were it is odd, in view of the excellent natural harbor and relatively easy terrain of Barrington, that there was not more mention of collection of tortoises by whalers and buccaneers a century or two ago. The *Academy* expedition found old bones — pelvises and limbs — estimated to have come from 14 individuals, but no shells. Some of the bones were from very large tortoises, and some from small but apparently

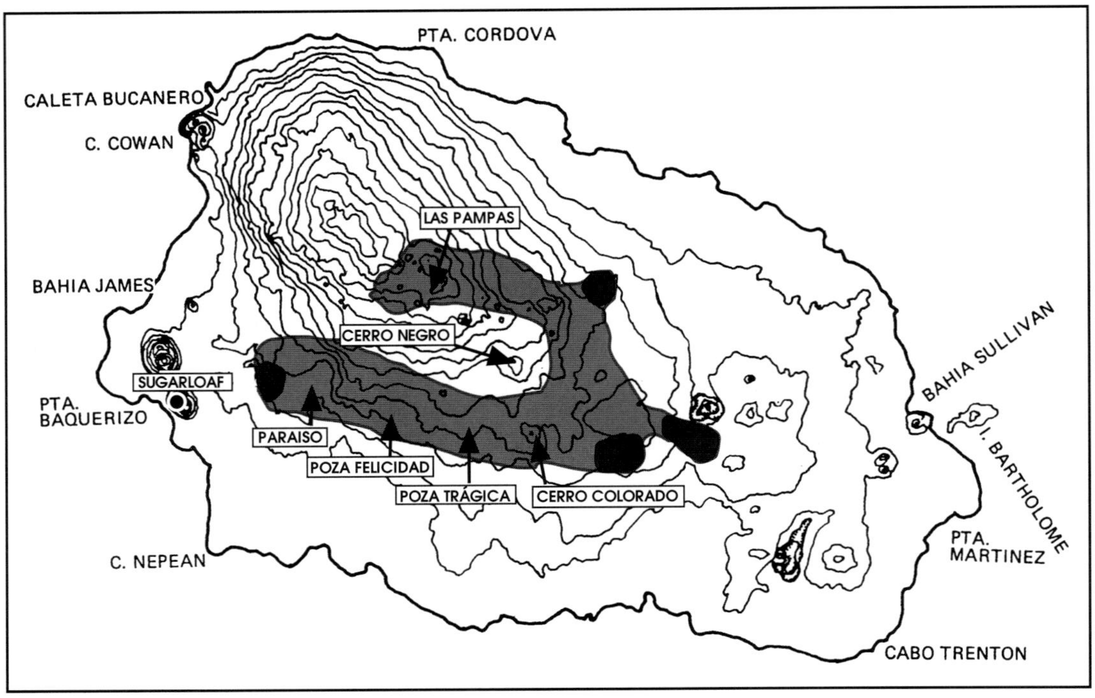

Map 9. James Island (San Salvador). Lighter shaded area is the distribution of tortoises, darker shaded areas are nesting zones. The dot represents an outlying record.

fully mature ones. As Van Denburgh said, "It would thus appear that the tortoises of Barrington Island varied considerably in size."

The disappearance of a native tortoise population from an island the size of Barrington, never inhabited by man, and without introduced predators (the only introduced mammals were goats, now extirpated) seems unlikely, in view of the persistence of small populations of tortoises elsewhere in the archipelago in the face of greater, multiple stresses. An alternative explanation is that the few commentators who reported tortoises from Barrington were either unreliable or had confused Barrington with another island. The nomenclature of the Galápagos Islands over the centuries has been exceedingly confusing (Woram, 1989), and it is significant, for example, that the Whalers' logbooks studied by Townsend (1925) made regular mention of tortoises collected on Crossman Island, a satellite of Albemarle far too small to have a native tortoise population.

As for Van Denburgh's account of the bony fragments found on Barrington, it is significant that no shells (not even fragmentary ones) were found, even though the shell is often the most durable component of a dead tortoise, and that the limb bones seemed to correspond to mature individuals of very divergent sizes. A possible explanation is that the excellent anchorage on Barrington encouraged overnight camping by whalers and other mariners, who may have taken ashore parts of butchered tortoises (i.e., entire limbs) from their on-board stocks of tortoises collected on various other islands, to serve as camp victuals.

Eibl-Eibesfeldt (1959) reported that some tortoises had been released on Barrington in 1956, and he had seen one of these, in good health, in the summer of 1957. There is a skull of a tortoise, collected on Barrington Island by Tjitte de Vries, in the collection of the Charles Darwin Research Station. Cayot (*pers. comm.*) reports that a large tortoise was found dead on Barrington in the 1980s, and that a live one had appeared at the visitor site as recently as 1988.

James (San Salvador or Santiago). — James is a large island, and today the tortoises are restricted to the far interior. However, this may not always have been the case. McBirney and Williams (1969) presented a detailed map of lava flows on James Island, and indicated that the coastal lowlands along about two-thirds of the shoreline, including the entire southern coastal zone, were composed of fresh, unvegetated lava. The western sections of these lava flows were already in place on Darwin's 1835 visit, but, in that the flow engulfed ceramic marmalade pots stashed by buccaneers in 1683 (Heyerdahl, 1963), it must be concluded that the major flows occurred between 1683 and 1835 (Simkin, 1984). Actually, major flows may have occurred only a short time before Darwin's visit. Porter (1815) collected no fewer than 14 tons of tortoises (about 500, averaging about 60 pounds) near a bay in the northeast, observing that only three of the tortoises were [adult] males, and all the females collected were, without exception, full of eggs, of which 10–14 were "hard" (i.e., shelled and ready for oviposition). The only bay of any consequence in northeastern James Island is Sullivan Bay, which today is in the midst of the enormous wasteland of fresh lava. No tortoises are found anywhere near the bay today. On a hike inland from Sullivan Bay in 1973, I walked about 40% of the width of the island before encountering a single adult female tortoise (see eastern nesting zone in Map 9).

The young Charles Darwin, on an overnight visit to the interior of James in 1835, found a party of Spanish tortoise hunters encamped six miles inland, at an altitude of nearly 2000 feet. During his visit Darwin subsisted entirely upon tortoise meat (Darwin, 1845). During this same period, Townsend (1925b) reported that whalers regularly obtained tortoises from James. The *Hector* obtained 23+ in 1834 and 124 in 1835. The *George and Susan* collected 68 and the *Benezet* 35+ in 1835. In 1836 the *Hesper* took 13 "plus many" and the *Lima* obtained 118 in 1836 and 224 in 1837. The *Phoenix* collected "12 plus seven boatloads" in 1838 and the *Chili* 93 in 1841. The *Equator* took 20 in 1845, but subsequent logbooks up to 1868 indicate no further tortoises collected on James.

By 1905–06 the tortoises were rare, even though the feral pigs that subsequently plagued the island had not yet been introduced (the date of introduction was estimated as "before 1930" by Hoeck [1984]). Goats had already been there a long time, although the precise history of their presence remains conjectural. It is reported that goats were introduced to James by Captain Porter in 1812, and although there is no reason to doubt this, these may not have been the goats that produced the current population (L. Cayot, *pers. comm.*). Ultimately the goats reached a population estimated at 100,000 (Calvopina, in Hoeck, 1984), but the *Academy* expedition (1905–06) never mentioned goats on James, and even in the 1980s the goats were reported to be progressively moving into new habitat, which would surely not have been the case if they had been on the island for nearly two centuries.

In 1906 Beck saw remains of the Spanish tortoise hunters' camp, still littered with tortoise bones, but found only 5 live tortoises (3 male, 2 female).

A salt mine was in operation at James Bay from about 1924 to 1930, and the workers apparently conducted an extensive tortoise oil business in the western and southwestern parts of the interior of James (A. Rambech, *pers. comm.* in MacFarland et al., 1974a). Long after the mine went bankrupt, heavy equipment remained on the island, and a caretaker, Sr. Leonardo Apolo, was still present as late as 1970. Apolo had a sympathetic interest in the tortoises and other fauna, and Leveque (1963) reported that a party from the Darwin Station, escorted by Apolo, had found two adult male tortoises, about 4 km inland, in April 1963. Apolo showed me what may have been the same individuals in 1970. At that time I also saw a small but old female tortoise entrapped in a steep-walled coastal tuff cone south of James Bay (Fig. 26), and found the skeleton of another female inland from James Bay.

Later in 1970, a major expedition to determine the status of the James Island tortoise was undertaken (Villa et al., 1971). These authors marked a total of 196 tortoises (129 males, 64 females, and 3 juveniles), and re-encountered all but 29 of 153 animals marked on a previous expedition. From these data they concluded that the population could not number more than 500 animals. Moreover, in all four subregions males outnumbered females and there was an extreme shortage of juveniles. Abundant nests were found distributed over four distinct nesting zones, and the condition of the eggs was generally excellent. But the presence of very large numbers of feral pigs resulted in the destruction of essentially all hatchlings. This observation is of interest in that, in the nesting zone of Indefatigable Island, feral pigs are primarily a menace to the eggs rather than the hatchlings, and a key conservation measure has been to erect heavy rock cairs over known nests, with interstices sufficiently wide to allow emerging hatchlings to escape. Villa et al. (1971) estimated that recruitment in the James Island tortoise population had started to decline about 50 years earlier, and had reached zero about 20 years after that. Villa's party killed about 1500 pigs during the four month stay, but they estimated that twice that number survived.

Figure 46. Harsh terrain on the northern slopes of Volcan Wolf near Punta Albemarle, Albemarle Island.

In 1982, Enriquez (1983) spent twelve months on James Island, and observed that pigs continued to be a threat to tortoise nests, and also that the native hawks (especially abundant on James) occasionally preyed upon hatchling tortoises. Tortoises were concentrated in the highlands from August through January, but from March until mid-May, the adults (males as well as females) were concentrated in the lower, nesting zones. (This contrasts interestingly with Porter's (1815) observation in the nesting grounds of northeastern James Island, where only females and subadult tortoises were found; the three males captured were found much further into the interior). Three shells of dead tortoises were found.

Márquez et al. (1992) reported that 158 young tortoises had been repatriated from eggs collected on James Island — 50 from 1970–71, 16 from 1971–72, 15 from 1972–73, 20 from 1973–74, 14 from 1974–75, 12 from 1975–76, 18 from 1976–77, 10 from 1981–82, and 3 from 1982–83. In addition, a total of 49 tortoises from subsequent years were still in captivity.

Albemarle (Isabela): Volcan Wolf. — The early history of the tortoises of this precipitous, arid mountain is obscure. The accounts of tortoise collection on Albemarle by nineteenth century whalers compiled by Townsend (1925b) did not differentiate between the various Albemarle populations, and it is probable that the majority were collected either in the humid areas in the south, with their enormous tortoise populations, or from Volcan Darwin, with the excellent anchorage offered by Tagus Cove. The lack of fresh water on Volcan Wolf prevented any settlement or agriculture and, while tortoises are found quite close to the sea at both Punta Albemarle to the north and, to a lesser extent, at Bank's Bay to the southwest, the former is an unstable anchorage and both present an extraordinarily rugged littoral zone that must be traversed with great difficulty before tortoise areas are reached (Fig. 46).

Thus, it was not until the twentieth century that the Volcan Wolf tortoises were documented. They may never have been abundant — there were neither feral mammals nor a history of collection by man when Beck collected the type specimen in 1901 and five others, also for Rothschild, in 1902 (Rothschild, 1901, 1903). Beck also collected for Van Denburgh in April 1906. Going ashore at Bank's Bay, this party found seven individuals after a week's search, but concluded that, in view of the dry weather, most of the tortoises had gone higher. None were found in a coastal wooded area north of Bank's Bay.

Subsequent survey data are few. Dowling (1962) found several young individuals at an altitude of about 800 feet, and Snow (1964), with two companions, found twelve in a single day's search. They were found at altitudes of between 230 and 600 feet, one being a juvenile (22.5 cm) and the remainder between 54 and 83 cm in carapace length. Snow also noted the presence of feral cats.

Snow (1964) also commented that, although the population probably numbered many hundreds of individuals, they occurred so close to the sea at Punta Albemarle that this was one of the chief areas for the illegal collection of live tortoises for export to the continent. This supposition is supported by the morphology of many of the confiscated live tortoises of unknown origin housed at the Charles Darwin Research Station; many of these show the broad, reverted anterior aspect characteristic of the Volcan Wolf tortoises.

My own observations are similar to those of Snow. In trips of a few hours each to Bank's Bay in 1971 and to Punta Albemarle in both 1972 and 1989, I found small numbers of tortoises on each occasion. The seven found near Bank's Bay on 21 January 1971 were all small to medium in size

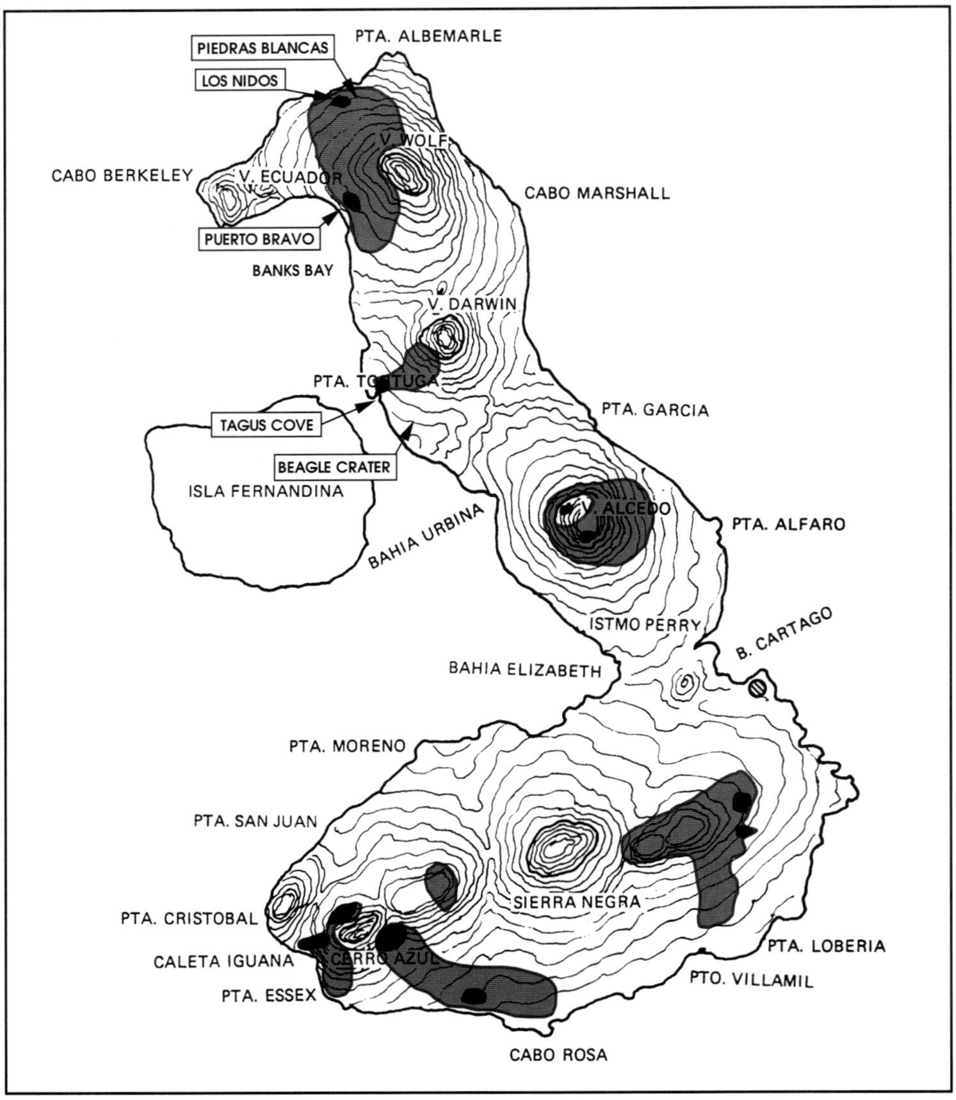

Map 10. Albemarle Island (Isabela). Lighter shaded areas are the distributions of various populations of tortoises, darker shaded areas are nesting zones. Striped area is the approximate former distribution of the extinct population reported by De Sola (1930), now under lava.

(Pritchard, 1971b), and included saddlebacked as well as dome-shelled individuals. At Punta Albemarle, I only found one in 1989, at an altitude of over 1000 feet. In the early 1970s I also found several old skeletons of animals whose plastra had clearly been separated by human action, so these tortoises have been utilized by settlers for food until quite recent times. Somewhat more recently, no fewer than 27 carcasses of female tortoises were discovered near Punta Albemarle (Anonymous, 1980). They had presumably been slaughtered when they came down to the coast to lay their eggs.

From 1980 to the present time, Volcan Wolf has been little visited by scientific parties, although there has been increasing illegal visitation and camping by fishermen and especially by sea cucumber (*pepino*) collectors. Cayot and Louis (1995) reported that six field trip reports were on file for the last fifteen years for both the Piedras Blancas and Puerto Bravo areas. In four of those from Piedras Blancas slaughtered tortoises were reported, as they were in all six reports from Puerto Bravo.

The total size of the Volcan Wolf tortoise population remains conjectural. Snow's (1964) estimate of many hundreds may have been too high — he assumed that the density he observed in an area of about 1 sq. km was sustained over a very large area, but this may not be the case. Nevertheless, the higher regions of Volcan Wolf are almost unexplored, and although much is fresh lava or very steep terrain, there may be many tortoises up there, and the relative abundance of juvenile specimens at the lower altitudes is encouraging. Occasionally, eggs have been collected and young tortoises head-started and repatriated — 14 from 1969–70, 22 from 1978–79, and 1 from 1979–80 (Márquez et al., 1992).

Figure 47. View of Volcan Darwin (Albemarle Island) from the eastern foot of Tagus Cove Mountain, February 1975. The closest area of forest, beyond the lava flow, is an important tortoise habitat and breeding area.

Unfortunately, significant slaughter of Volcan Wolf tortoises continues up to the present time. Louis (1994) reported on a visit to Piedras Blancas and Los Nidos, on the northern side of Volcan Wolf, on 11–15 August 1994, when he found 10 live and 12 dead tortoises. He commented on the extremely easy access to the area and the close proximity of illegal fishing boats. At another site, Puerto Bravo, west-southwest of Volcan Wolf, 18 live and 12 dead tortoises were found. Some of the carapaces were found at an illegal camp occupied a couple of months earlier by sea cucumber fishermen.

Albemarle (Isabela): Volcan Darwin. — This mountain provided the tortoises that, for centuries, have been removed through the excellent deep-water anchorage at Tagus Cove. The intensity of this collection during the nineteenth century

Figure 48. Volcan Darwin (Albemarle Island) tortoise resting at the western foot of the mountain. Note the land iguana (*Conolophus subcristatus*) living in microsympatry with the tortoise. April 1972.

Figure 49 (left). Adult male tortoises in an area of the Volcan Alcedo, Albemarle Island, caldera rim dominated by tree ferns (*Cyathea weatherbyiana*). July 1986.

Figure 50 (upper right). Post-hatchling tortoise (carapace length ca. 10 cm) entering a shallow burrow or retreat. Inner slope of the southern flank of Volcan Alcedo. April 1985.

Figure 51 (lower right). Adult female tortoises in a well-grazed area on the eastern rim of Volcan Alcedo. August 1983.

is unclear, in that the whalers' records do not differentiate between the different parts of the very large island of Albemarle, but Beck (1903) was probably correct when he observed that the Volcan Darwin tortoise ("*Testudo microphes*" [sic]) had been used as food for whalers more than any other. Beck found large numbers of tortoise trails, indicating a former population of hundreds of tortoises, but, he reported, "today it is a hard matter to find one." But on a later trip (March – April 1906), Beck's party collected 13 males and a young female. His designation of the locality as Tagus Cove is confusing, in that this drowned crater and associated coastal hill represent the access route rather than the collecting site. The tortoises were collected at the foot of Volcan Darwin itself, across a zone of rough lava that would be impassable to tortoises (Figs. 47–48). Occasionally a tortoise is found near Tagus Cove itself, but there does not appear to be a population there, and the origin of such individuals is unclear.

Twentieth century authors including Beebe (1924), Eibl-Eibesfeldt (1959), and Bowman (1960) found no live tortoises at Tagus Cove at the western foot of Volcan Darwin, although their discovery of dead young tortoises did indicate that reproduction was occurring. Snow (1964) climbed to an altitude of 3000 feet on western Volcan Darwin, finding fresh tortoise droppings and trails but no actual animals. But MacFarland et al. (1974a) reported an estimated population of 500–1000 animals. A total of 65 tortoises had been marked in October 1965, and the size and sex distribution of these (6 adult males, 26 adult females, 2 young males, 1 young female, 8 non-sexed, 22 small) was encouraging. This is an arid area, and much of western Volcan Darwin as well as most of the western lowlands are unvegetated fresh lava, unsuitable for tortoises. The population apparently descends to a small area of the western lowlands during rainy periods, but otherwise is dispersed high up on the mountain. On an ascent of Volcan Darwin on 22 January 1971, when it was very dry, I reached the lowland tortoise zone within an hour, but found only old droppings and two crumbling skeletons. A single juvenile tortoise, about 40 cm in length, was found half way up the

Figure 52. Adult female tortoise from the Volcan Alcedo caldera floor, showing extensive carapacial scarring, possibly as a result of contact with hot volcanic ejecta. August 1983.

mountain, and five larger ones, including an adult female, were found close together near the caldera rim (Pritchard, 1971b). Yet on a later trip, in 1972, I encountered 17 tortoises of a wide range of sizes in the small vegetated lowland area.

This population would seem to have good prospects, unless the feral cats become a problem for the hatchlings. But it is likely to be limited from reaching high density by the low rainfall, rugged terrain, and relatively modest extent of the area suitable for tortoises. Louis (1994) visited the Tagus Cove area from 16–18 August 1994, and was able to collect blood samples from 25 live tortoises. He found no dead tortoises, but did observe that goats, previously considered to be restricted to the southern parts of Isabela, had reached the area, the rim of Beagle Crater being carpeted with goat droppings. One tortoise was found killed at Urbina Bay (Volcan Darwin) in 1984 (Cayot and Louis, 1994).

Albemarle (Isabela): Volcan Alcedo. — The nineteenth century literature makes no mention of the tortoises of Volcan Alcedo, and it is probable that they occurred too far inland to be convenient victualling for whalers. Moreover the shoreline on the eastern side of the mountain offers no good anchorage. Nevertheless, Van Denburgh (1914) reported that Beck's party had encountered remains of a camp on the upper eastern slopes of Alcedo, at 2200 to 2500 feet elevation, where the remains of about 70 old tortoise skeletons indicated that the tortoises had been subjected to heavy exploitation. The camp was presumably long abandoned and the tortoise shells already very deteriorated, in that no collectible carapaces were found. But the feral burros that had been used to transport kegs of tortoise oil to the coast and that now, as in 1905–06, run free on this mountain are presumably a permanent legacy of the operation.

Figure 53. Group of subadult tortoises in a pond (since dried up) beside the fumarole, inner slope of southwestern part of Volcan Alcedo caldera, Albemarle Island. March 1970.

Figure 54. Tortoises in the mist. Concentration of mostly adult tortoises in a cloud-catchment drip zone on southern rim of Volcan Alcedo. July 1986. Also see figure and poem in the editorial prologue.

Beck's party found and collected a single live female tortoise. This specimen had mature ovaries but had a carapace length of only 67.75 cm. It was very steeply domed and may have been slightly kyphotic — certainly it was an unfortunate specimen on which to base conclusions about the morphology of the population.

Beck's report of an essentially vanished tortoise population, and grass so high and thick that ascent to the rim of the caldera was all but impossible, apparently discouraged others for decades to come. But in August 1963, M. Castro was able to climb Alcedo from the northeast, reaching the rim and spending two nights high up on the mountain (Snow, 1964). He located an apparent tortoise nesting area 5 km north of the crater rim and found many old tortoise droppings, but saw no live animals. The following year, in February 1964, members of the Galápagos International Scientific Project, flying by helicopter over the volcano, saw many tortoises feeding in the grassy areas of the caldera floor. Finally, Eric Shipton (the renowned Himalayan mountaineer) reached this area by foot in 1965, and was able to confirm the existence of a very large tortoise population (de Vries, 1984).

In the following years, a route was developed whereby the crater rim and floor could be reached without great difficulty, and subsequently, numerous parties — even relatively large groups of tourists — have ascended Alcedo and observed the largest tortoise population in the Galápagos today, and the only one in a semblance of primordial abundance (Figs. 5, 49–55). The population was estimated at 3000–5000 individuals by MacFarland et al. (1974), although Fowler (1983) observed that the methodology used by these authors was unstated, and that the actual population was probably closer to the lower figure. Various aspects of this population have been studied in depth by those who have managed to solve the water supply problem; Fowler (1983), for example, camped in the crater for over a year while studying the feeding ecology of the tortoises and the feral burros.

Today the tortoise population is certainly large and includes a satisfactory proportion of young animals (indicating good reproduction) intermingled with some of the largest tortoises in the world today (indicating good adult longevity). Human exploitation has not been significant since the early part of this century. Cayot and Louis (1994) reported that only three slaughtered tortoises were found on Alcedo during the 1970s, and these were estimated to have been killed during the preceding decade. Nevertheless, the presence of feral burros, cats, and rats underscores the need for careful monitoring. The impact of the first of these has been studied by Fowler (1983), Fowler and Roe (1984), and Fowler and Johnson (1985). These studies show that 18.2% of 88 nests in the caldera floor were disturbed by

Figure 55. More tortoises in the mist. A group of mostly adult tortoises on the rim of Volcan Alcedo, August 1986.

the trampling of burros. The percentage was higher on the southern nesting site (32.1%) than on the northern (11.7%).

In addition to these problems, in recent years goats have succeeded in crossing Perry Isthmus and are now widespread on Alcedo. The threat that they pose to the tortoises was graphically described by Merlen (1992). Muñoz (1993) reported that goats had colonized Alcedo within the last four years. A study was conducted in 1991 to determine their current distribution, population structure, home range, reproductive cycles, and feeding behavior. The goats were found on the southern, eastern, and southwestern flanks of the volcano, with three groups averaging 18.2 individuals each within the caldera itself. Seasonal movements were found, as the goats migrated northwards or sought food or water during the *garua* season. By 1995, the goat population had increased catastrophically, to 20–30,000 individuals, and the tortoise population must now be listed as threatened (L. Cayot, *pers. comm.*). Volcan Alcedo is now closed to tourism, and the Charles Darwin Research Station and the Galápagos National Park Service now plan a major goat killing campaign (Cayot and Snell, 1996b).

Volcan Alcedo has permanent hot fumaroles, sulfur vents, and other hot spots, and these may occasionally kill tortoises. Individuals are sometimes seen with heavy scarring of the dorsal scutes (Fig. 52), possibly as a result of exposure to volcanic materials. Hayes and Beaman (1985) found a deformed adult female tortoise that, they speculated, may have been a survivor of an early burro injury, although other causes of the deformity, including genetic or teratogenic ones, were also considered.

It is important to note that, although very high densities of tortoises can be seen on Alcedo, the actual area of suitable tortoise habitat is not particularly extensive. Tortoises may be found on most of the rim of the caldera (an essentially linear distribution) and are found in scattered fashion on the upper eastern flanks of the volcano. But much of the caldera floor is precipitous lava, the lower eastern slopes are largely almost unvegetated pumice, and the entire western side of the volcano is densely vegetated, interlaced with vertical-sided ravines, and has very few tortoises. The highest numbers are found in the flat or rolling, grassy or lightly wooded areas below the fumarole on the southern side of the caldera floor (Fig. 53). Population movements are extensive. During dry times, the tortoises concentrate along areas of the caldera rim where moisture from passing clouds is entrapped by low trees and forms heavy morning dew or even puddles (Figs. 54–55; see also figure and poem in the editorial prologue). But during wetter times, the tortoises migrate to lower altitudes, many of them concentrating in lush grassy meadows on the caldera floor, where seasonal pools may attract particulary large numbers.

Albemarle (Isabela): Sierra Negra and Cerro Azul. — Originally, this area of southern Albemarle must have harbored enormous populations of tortoises, judging by the great areas of what appear to this day to be suitable habitat. But the populations are severely reduced and tortoises have been totally extirpated from vast areas on the southern uplands of Sierra Negra where they once abounded.

The impact of the whalers on this area is somewhat conjectural, in that few of the whalers specified the part of Albemarle from which their tortoises were obtained (see back cover engraving, with figure caption on p. 4). In some cases the reports indicate not the main island of Albemarle, but satellite islets far too small to have tortoise populations (e.g., Cowley Islet, or Crossman). It can probably be assumed that most of the 2493 tortoises collected from Albemarle by whalers whose logbooks were examined by Townsend (1925b) were collected either from Tagus Cove in the north, or from Villamil or Iguana Cove in the south; but the actual distribution of the collecting effort will never be known for sure.

The tortoises were still abundant in 1902 (Fig. 56), but their prospects were bleak, when Beck spent two weeks at a newly-established ranch ten miles from the shore on southern Albemarle, presumably on Sierra Negra. Cattle roamed the mountain in great numbers, as did feral dogs, and while the former may not have presented serious problems to the tortoises, the dogs were reported to eat young tortoises as fast as they were hatched, and Beck doubted if more than one in 10,000 escaped. The dogs even killed adult tortoises, primarily the smaller females, but occasionally even large males.

As bad as the dogs, or even worse, were the 50 or so field hands employed on the ranch, who killed tortoises not only for food, but also for their oil — a valuable export commodity. Although the proprietor of the ranch claimed that only male tortoises were killed, in fact females as well as males were slaughtered without restraint, and sometimes only a few pounds of meat was removed from an adult tortoise. Beck found about 150 tortoise skeletons at a single waterhole (Fig. 57), and about 100 at another. He found 7 tortoises in a pond 10 miles from the ranch (Fig. 56), but feared that they were doomed within the coming months.

On a return trip only four years later, Beck was able to review the impact of the massacre (Fig. 57). He was able to collect about 40 tortoises, but nearly all were females, and only 2 males more than 28 inches in carapace length were included. Hundreds of tortoise bones were found along the trails and at the water holes. The natives said that all tortoises had been killed off below the settlement and that they were no longer abundant anywhere on the mountain.

Settlements on southern Albemarle have continued through this century, including a major penal colony in the interior from 1946–59, a military presence during WWII, and the existing coastal village of Villamil, founded in 1895, connected to the highland community of Santo Tomás by foot trail. In 1957, young tortoises could still be purchased from Villamil residents (Eibl-Eibesfeldt, 1959). Exploitation of the tortoises by settlers continued into the 1960s

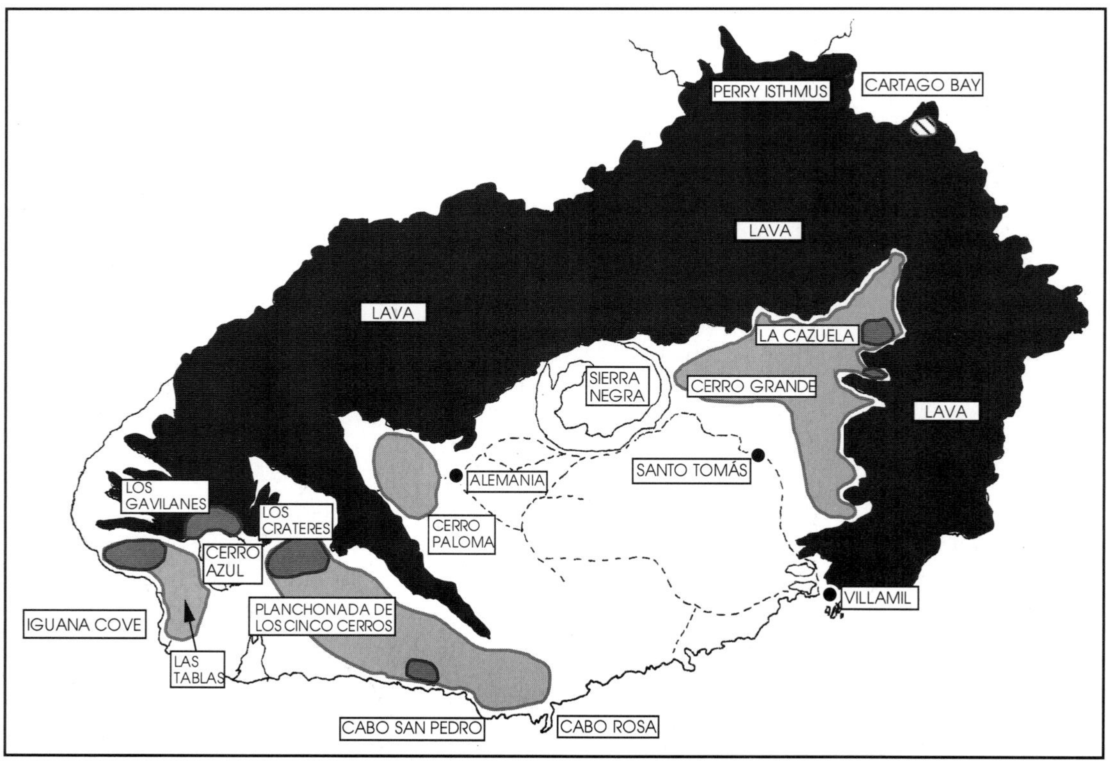

Map 11. Southern Albemarle Island (Isabela): Sierra Negra and Cerro Azul. Named tortoise regions designated. Lighter shaded areas are the distributions of various populations of tortoises, darker shaded areas are nesting zones (*galapagueras*). Striped area is the approximate former distribution of the extinct population reported by De Sola (1930), now under lava.

Figure 56. Scene on southern uplands of Sierra Negra, Albemarle Island, in 1901. Tortoises on this mountain are now very few and dispersed. Photo by R.H. Beck, reprinted from Beck (1903) and Slevin (1959). The 1959 caption read: "A typical waterhole scene of 1901 on top of Villamil Mountain, Albemarle Island, the site of probably the largest tortoise colony on the islands until the oil hunters began operations."

(Snow, 1964), and even later. MacFarland et al. (1974a) reported that settlers still killed tortoises on Sierra Negra, and that, as recently as 1971, remains of 60 slaughtered tortoises had been found in an area of 0.9 sq. km, although only 8 appeared to have been slaughtered recently, the remainder 2–5 years earlier.

MacFarland et al. (1974a) also documented the virtual extirpation of two of the pockets of tortoise abundance on southern Albemarle that had survived the onslaughts of the settlers and oil hunters. Tortoises had been almost completely eliminated from 120 sq. km of the uplands of Sierra Negra by oil hunters from the prison colony in the 1940s and 50s, with operations extending even to Cerro Azul. And, between 1959 to 1969, tortoises were decimated in an area of about 5 sq. km inland from Iguana Cove, in the extreme southwest, by employees of cattle companies. The latter area had previously been an important tortoise area, and had provided the vast majority of the 180 mostly juvenile tortoises collected by Townsend's party during three weeks in 1931. However, L. Cayot (*pers. comm.*) reports that Iguana Cove still had a good tortoise population in 1995.

Tortoises survive in low density in three metapopulations on southern Albemarle. An estimated 300 individuals survive in a dry area near Cerro Grande, east of the caldera of Sierra Negra and inland from Cerro Ballena. Most tortoises found here are young, suggesting good natural reproduction and recruitment. Villa (1972) reported that 217 individuals of this population had been marked. This colony is widely separated by settlements and cultivation from a lowland southern population of perhaps 200 individuals inland from Cabo Rosa. And around Cerro Azul, tortoises may be found from low altitudes (the humid zone extending down to sea level on this mountain) all the way to the caldera rim. This population is estimated at 500–700 individuals, of which 203 had been marked by 1971 (Villa, 1972). A few individuals also survive on the eastern side of a long lava flow east of Cerro Azul (Fig. 58). This flow reportedly occurred in 1926 (J. Gordillo, *pers. comm.*), nearly reaching the sea, and dividing the Cerro Azul tortoise population.

The Cerro Azul population utilizes four well-defined nesting zones: *Las Tablas* (west of Cerro Azul), *Los Crateres* or *Pegas* (east of Cerro Azul), *Planchonada de los Cinco Cerros* (west of San Pedro), and *Los Gavilanes* northwest of Cerro Azul (Villa, 1972). In most of these localities, pressure from feral mammals (pigs, dogs, and possibly cats) was intense, pigs raiding the nests and dogs killing young and even half-grown tortoises up to 55 cm in curved carapace

Figure 57. Scene on Sierra Negra, probably in 1902. Photo by R.H. Beck, reprinted from Beck (1903). The original caption read: "Work of the oil hunters."

length. Feral cattle were also abundant and constituted an important food source for the dogs.

Conservation efforts in recent years have concentrated on educating the local population, field patrols and law enforcement, control of feral dogs by poisoning (Moore and Moore, 1982), protection of natural nests from predation by erecting rock cairns, and sporadic head-starting of eggs from wild nests. Márquez et al. (1991) reported 25 tortoises repatriated to Cerro Azul from 1971–72; 25 from 1972–73; 14 from 1973–74; 37 from 1975–76; 17 from 1977–78; 16 from 1978–79; 17 from 1979–1980; 14 from 1980–81; 14 from 1981–82; and 3 from 1986–87; and 33 more were still held in captivity. On Sierra Negra, only 13 young tortoises had been repatriated from artificially incubated eggs laid between 1977 and 1982, although 53 animals from 1983–84 and subsequent seasons were still held in captivity.

The wide hiatus between the highland tortoise populations southeast of Sierra Negra and east of Cerro Azul is anthropogenic in nature, the result of the intensive slaughter of tortoises and conversion of native vegetation to pasturelands as described by Beck (1903). On the other hand, recent maps of the geology of southern Albemarle show virtually continuous lava flows over the entire northern half of this enormous block of land, reaching from (and joining) the two calderas, and extending north to engulf Perry Isthmus. No tortoises could survive in this lava field today, but there was a populous colony in the lowlands of the eastern side of Perry Isthmus until comparatively recent times. Townsend (1925) reported that a tortoise of extraordinary size, famous among visiting whalers, lived in this area, and was seen by George A. Grant of Nantucket in 1881. The animal was known as *Port Royal Tom*, and had dates and names carved into its carapace, the earliest date being 1791. It was in this same area that De Sola (1930) observed numerous tortoises, some of which he collected alive, and which he named *Testudo vandenburghi*. McBirney and Williams (1969) reported three lava flows since 1930 from vents not far from the northern rim of the Sierra Negra caldera, in 1948, 1953–54, and 1963–64. The last flowed primarily towards Punta Moreno and the western side of Perry Isthmus, but either of the two earlier ones may have destroyed the tortoise habitat on the eastern side of the Isthmus.

A new threat flared up in February 1985, when a great fire, apparently started by settlers, burned a large area of southern Albemarle. Tortoises were rounded up and preparations were made to evacuate them by helicopter, although this turned out to be unnecessary. An international firefighting effort was mounted and eventually a firebreak line 40 km long was cleared. Tortoises were regularly encountered along this firebreak, but none was reported killed by the fire, although 175 sq. km of wilderness was devastated, and two subadults died from the extreme heat while being carried to a corral prepared for them in a coastal area near San Pedro. The fire was not finally extinguished until rains came in July (Márquez, 1986; Nowak and Lawesson, 1988). An unfortunate side-effect of this fire was the new access it created for poachers to reach, and to kill, tortoises (Wallner, 1995). Another was that an aggressive exotic plant, the *guayabo*, rapidly colonized areas where native vegetation had been destroyed.

Because of their severely reduced populations and ongoing stresses, tortoises from the Cerro Paloma and Cazuela populations (west and east of the Sierra Negra caldera, respectively), were transferred to a captive breeding facility, or *crianza*, at Villamil. Currently, one adult female, five adult males, and five juveniles from Cerro Paloma reside at this facility, as well as approximately 75 tortoises from Cazuela (only 15 adults). There are still five adult males (no known adult females) in the wild at Cerro Paloma, and a few wild tortoises at Cazuela (L. Cayot, *pers. comm.*; Cayot and Snell, 1996a).

The most recent summary of the status of southern Albemarle tortoise populations is that of Cayot and Márquez (1994). These authors estimated that there were 1000–2000 tortoises on Cerro Azul, with subpopulations of 100–300 at Las Tablas, 500-1000 at Los Crateres, less than 100 at Los Gavilanes, and 100–300 at Cinco Cerros. Tortoises were considerably scarcer (less than 500) on Sierra Negra, with an estimated 100–200 at Cabo Rosa, less than 100 at Roca Union, less than 20 at Cerro Paloma, and less than 100 at Cazuela.

Recent reports indicate that problems for the tortoise populations of southern Isabela are far from over. A second fire, apparently started when a camp cooking fire was incompletely extinguished, occurred in April 1994 in the southwestern sector of Sierra Negra (Cruz et al., 1994). The fire was not nearly as disastrous as the 1985 fire, but it was not extinguished for almost two months. Furthermore, Cruz Márquez and colleagues were able to travel with the helicopter assigned to assist with fire control, and although they found no tortoises killed by fire in the nearby Roca Union subpopulation, they found the remains of eight tortoises (3 males, 3 females, 2 juveniles) that had clearly been killed by man for their meat (Márquez et al., 1994). At Cerro Paloma, remains of further tortoises were found, but these appeared to have been killed before 1988.

Southern Albemarle is an enormous area, conservation staff are few, comprehensive patrols are almost impossible, and the Ecuadorian settlers are resentful of increasingly restrictive conservation regulations. In 1986, the Galápagos Marine Resources Reserve was established, and in August 1992 a marine management plan for Galápagos waters was signed into law (Camhi, 1994). The regulations were de-

Figure 58. Adult male southern Albemarle tortoise, east of Cerro Azul. March 1971.

signed to restrict foreign, industrial scale fisheries in Galápagos waters but to make concessions to the economic needs of local fishers. Nevertheless, the settlers on southern Albemarle have been vociferous in their objections to the fishery regulations and, on and off since July 1993, they have protested and picketed the Charles Darwin Research Station, threatening to kill giant tortoises (even *Lonesome George*, the last survivor of the Abingdon Island population), and to introduce non-native species to the few remaining pristine islands. In September 1995, the Park Headquarters and the Darwin Station were occupied for two weeks by local protesters (Lemonick, 1995; Merlen, 1995).

Slaughtered tortoises were indeed found, although in places where there is no certainty that the slaughter had been done by the same individuals who had made the threats, or even by fishermen at all. Over 80 tortoise mortalities were confirmed on Albemarle alone in 1994 (Camhi, 1994; Cayot and Louis, 1994). The government capitulated, rescinding the ban on large-scale commercial fishing of sharks, sea cucumbers, lobsters, and groupers. Louis (1994) documented a number of the tortoise mortalities with photographs, and reported on the tortoises he found on southern Albemarle between 20 July and 30 August 1994, as follows:

1) Roca Union. 22 July. Twenty-one live tortoises found, as well as 7 previously reported carcasses and 5 unreported ones. Plastron had been removed by chopping in all 12 cases.

2) Cazuela. 23 July. Only one carapace found, but approximately 39 dead tortoises had been found here in March 1994.

3) Cinco Cerros. 24–25 July. Fourteen live tortoises found. Later (22–26 August) in the same area, 7 dead tortoises found close to the beach, and 4 more inland. In 8 of these cases, the tortoises had been butchered. Forty-six live tortoises found on the second visit.

4) Las Tablas. 28–30 July. Twenty-five live tortoises found, all females and juveniles.

5) Cerro Paloma, Los Crateres, and Los Gavilanes. 1–6 August. Five adult males found at Cerro Paloma, and about four dead tortoises. Twenty-five live tortoises found at Los Crateres, and 8 at Los Gavilanes.

Named tortoise areas or *galapagueras* on southern Albemarle are located as follows, in relationship to the two calderas (Cerro Azul in the west and Sierra Negra in the east):

1. La Cazuela is east of Sierra Negra, towards the coastal hill Cerro Ballena, and is surrounded on all sides but the west by lava flows.

2. Los Gavilanes is north and northwest of Cerro Azul.

3. Los Crateres is east and northeast of Cerro Azul.

4. Cerro Paloma is midway between the calderas of Cerro Azul and Sierra Negra, just east of the long tongue of unvegetated lava left by a 1926 eruption.

5. Las Tablas is southwest of Cerro Azul, extending from the caldera to the sea, and including Iguana Cove and Cerro Manzanilla.

6. Cinco Cerros is located southwest of Cerro Azul, extending from the caldera to the sea, and encompassing the coastal areas between Cabo San Pedro and the lava flow due south of Cerro Azul. The Cinco Cerros (= five hills) are isolated cones arranged in a northwest-southeast radial fashion between the caldera and Cabo San Pedro.

7. Cabo Rosa is located on the southern coast of southern Albemarle, in the lowlands opposite the saddle between Cerro Azul and Sierra Negra.

8. Roca Union is a rock island in the ocean due east of Cabo Rosa and due south of Santo Tomás; in the above accounts, the name refers to the lowlands of southern Albemarle directly facing Roca Union.

CONCLUSIONS

Neither the nomenclatural nor the survival status of Galápagos tortoises can be considered settled.

Nomenclatural Status. — In order to stabilize the nomenclature, it is necessary to make petition to the ICZN to suppress several names that could, under previous editions of the Code of Nomenclature, have been declared *nomina oblita* (e.g., *Testudo californiana* Quoy and Gaimard, 1824a), or that have been widely used in an incorrect context (e.g., *Testudo ephippium* Günther, 1875), or whose type localities, and thus whose islands of origin, are uncertain (e.g., *Testudo nigrita* Duméril and Bibron, 1835). Some new subspecies could potentially be described (e.g., on northwestern Indefatigable Island, or northeastern Chatham Island), but will require type material (presumably salvaged) before names can be proposed and descriptions drawn up.

Furthermore, a few recommendations can be made to develop a degree of sophistication in Galápagos tortoise nomenclature that is now lacking:

1) Morphological comparisons need to be made between the abundant museum material of Galápagos tortoises, collected in the nineteenth and early twentieth century, and the current (living) generation of tortoises on all islands where they still survive. This will generate invaluable insight into the possibility of microevolution occurring even within a few generations, as occurs with the geospizine finches in the archipelago.

2) Galápagos tortoises are long-lived but they are not immortal, and opportunities should not be lost for collecting any skeletal material in reasonable condition of tortoises that have died (or been killed) in the wild. Using Global Positioning Systems, the locality data of such specimens can be recorded with precision. Comprehensive osteological collections will not only facilitate the first recommendation above, but will counterbalance the strong bias towards entire, mounted specimens in extant museum collections.

3) The holotypes of the several currently indeterminate "species" of Galápagos tortoise (*T. californiana* and *T. nigra*, *T. elephantopus*, *T. nigrita*, *T. planiceps*, *T. microphyes*, *T. wallacei*, *T. clivosa*, and *T. typica*) need to be re-examined. In those cases where dried soft tissues are still present on the specimen, even in minuscule amounts, it is possible that mitochondrial DNA analysis and appropriate comparisons with specimens of known origin may allow these names to be allocated definitively. Genetic studies on all extant populations are currently being conducted by Edward Louis, and the results of these may well also have an influence upon nomenclature.

Survival Status. — Eibl-Eibesfeldt (1959) noted that about 2000 people lived in the Galápagos Islands, and concluded: "On the whole colonization conflicts with nature protection. How can both interests be reconciled and what are the prospects?" Today, almost 40 years later, we would have to agree with the statement, and confess that we have no definitive answers, but a generally pessimistic response, to the questions.

For a while things looked better, but today there is much bleak news to report about the survival prospects of the various populations of tortoises in the Galápagos. Despite 35 years of hard, dedicated work by the staff of the Charles Darwin Research Station and the Galápagos National Park Service, new problems keep arising, and feral mammals continue to reach, or be taken to, new areas. Tourism, initially the cash cow that, through tourist taxes, financed and justified the National Park and the various conservation programs in the archipelago, now is revealing its disadvantages. Flights to the islands now occur almost daily, and money is made by supplying tourists with accommodations, transport, guides, meals, and souvenirs. No one starves in the Galápagos nowadays, and the archipelago, despite its rugged terrain and reputation as a place of shipwreck and hardship at least in former years, now has the highest overall standard of living of any province of Ecuador.

This has prompted massive increases in the human population of the islands. Under the Constitution of the Republic of Ecuador, this migration cannot legally be stopped. Yet many of the immigrants find that they cannot live directly from the tourist trade, and are forced to extract a living from the Galápagos environment itself, by agriculture or fishing, to survive. Inevitably this leads to conflicts in an area where 95% of the land surface is National Park, and where there is also a Marine Conservation Zone around the islands. Of late these conflicts have become so severe that the very future of the Galápagos, its ecosystems, and its endemic species, is at stake.

Today, the habitat of the tortoises on Indefatigable overlaps substantially with private agricultural property, and stray dogs and feral pigs wander through the Tortoise Reserve. On Duncan Island, the tortoises have possibly been saved through the head-starting program, but rats still exist, and it is far too early to declare a victory and retreat. On Albemarle, tortoises are threatened by direct human action on all the volcanoes except Alcedo. Alcedo itself, home to the largest tortoise population in the archipelago, seemed secure until recently; but now, tens of thousands of goats have joined the long-established feral burros, and the vegetative cover of the mountain will be cleared within a few years unless an extraordinarily ambitious goat control program is mounted.

The tortoises of Hood Island may survive. Indeed, it is vastly encouraging to see a population of several hundred tortoises where none existed a few years ago. But one cannot ignore the possibility of inadequate genetic diversity in a population derived from just three males and a dozen females. Tortoises on James Island still have to compete with tens of thousands of goats and their eggs and young are still exposed to depredation by pigs. Perhaps on Chatham, now that the feral dogs have been controlled, the tortoises may be safe. But one can never relax and unremitting vigilance will be necessary. Major new educational programs will have to be launched, for both the young people and the adults of the Galápagos Islands. The possibility exists that captive breeding groups for all subspecies may be necessary, at least until the troubled sociological and political situation in the islands calms down. A very firm hand will have to be taken with those who breach park regulations, and especially with those who set up illegal camps, poach protected species, or deliberately or accidentally introduce new animal (and plant) species to pristine areas.

The tortoises of Charles, Barrington, Jervis, Narborough, Abingdon, and southwest Chatham Islands have been exterminated in a time span that, geologically speaking, is a blink of an eye. Surely their surviving relatives cannot be far behind unless drastic remedies are unleashed promptly.

LITERATURE CITED

ADLER, K.K. 1989. Herpetologists of the past. In: Adler, K.K. (Ed.). Contributions to the History of Herpetology. Soc. Study Rept. Amph.

AGASSIZ, L. 1857. Contributions to the Natural History of the United States, Vol. 1. Boston: Little, Brown, and Co.

ANONYMOUS. 1894. Editorial. Novitates Zoologicae 1:1.

ANONYMOUS. 1980. News from Academy Bay. Noticias de Galápagos 33:1-6.

ANONYMOUS. 1982. Is there a mate for Lonesome George? Noticias de Galápagos 35:2.

ANONYMOUS. 1986. The Española tortoises — a very special case. Noticias de Galápagos 44:5.

ATKINSON, G. 1922. The extraordinary voyage in French literature. Vol. 2. From 1700 to 1720. Paris, Champion.

AUFFENBERG, W. 1967. Notes on West Indian tortoises. Herpetologica 23:34-44.

AUFFENBERG, W. 1971. A new fossil tortoise, with remarks on the origin of South American testudinines. Copeia 1971:106-117.

AUFFENBERG, W. 1974. Checklist of fossil land tortoises (Testudinidae). Bull. Fla. State Mus., Biol. Sci. 18:121-251.

BACON, J.P. 1978. A tortoise goes home. Zoonooz 51(12):4-7.

BACON, J.P. 1980. Some observations on the captive management of Galápagos tortoises. In J. B. Murphy and J. T. Collins, eds., Reproductive Biology and Diseases of Captive Reptiles. SSAR, Contrib. in Herpetol. 1:97-114.

BAILEY, A.M. 1970. Narrative of the 1960 field trip of the Denver Museum of Natural History to the Galápagos Islands. Denver Mus. NMat. Hist., Museum Pictorial 19:1-86.

BANNING, G.H. 1933. Hancock expedition of 1933 to the Galápagos Islands, General Report. Bull. Zool. Soc. San Diego 10:1-30.

BARBOUR, T. 1914. On some Australasian reptiles. Proc. Biol. Soc. Washington 27:201-206.

BAUR, G. 1889. The gigantic land tortoises of the Galápagos islands. Amer. Nat. 23:1039-1057.

BEAMAN, K.A. 1985. Bibliography of *Geochelone elephantopus*. Smithsonian Herpetological Information Service 65:1-18.

BECK, R. 1903. In the home of the giant tortoise. 7th Ann. Rept. New York Zool. Soc. 7:160-174.

BEEBE, W. 1924. Galápagos: World's End. G.P. Putnam's Sons, New York.

BLACK, J. 1973. Galápagos: Archipielago del Ecuador. Imprenta Europa. Quito.

BOCOURT, F. 1868. Description de quelques chéloniens nouveaux appartenant à la faune Mexicaine. Ann. Sci. Natur. Zool., Paris (5)10:121-122.

BOULENGER, G.A. 1889. Catalogue of the Chelonians, Rhynchocephalians, and Crocodiles in the British Museum (Natural History). Taylor and Francis, London.

BOUR, R. 1980. Essai sur la taxinomie des Testudinidae actuels (Reptilia, Chelonii). Bull. Mus. Natn. Hist. Nat., Paris, (4)A, 2(2):541-546.

BOUR, R. 1985. Les tortues terrestres géantes des îles de l'océan indien. Studia Geologica Salmanticensia, Vol. Esp. 1. Studia Palaeocheloniologica I:17-76.

BOUR, R. 1988. Tortues et insularité: les tortues des Seychelles. Bull. Soc. Zool. France 112(3-4):401-418.

BOURGAT, R., AND BOUR, R. 1983. La tortue géante de Perpignan *Cheirogaster perpiniana* (Deperet, 1885). Soc. Agricole, Scientifique et Litteraire des Pyrenées-Orientales, Perpignan 91:167-177.

BOWMAN, R.I. 1960. Report on a biological reconnaissance of the Galápagos Islands. UNESCO, Paris, 65 pp.

BRAITHWAITE, C.J.R., TAYLOR, J.D., AND KENNEDY, W.J. 1973. The evolution of an atoll: the depositional and erosional history of Aldabra. Phil. Trans. Roy. Soc. Lond. B 266:307-340.

BROOM, R. 1929. On the extinct Galápagos tortoise that inhabited Charles Island. Zoologica 9(8):313-320.

CAMHI, M. 1994. Sharks in Galápagos in peril. Shark News (Newsletter of the IUCN Shark Specialist Group) 2:1-3.

CAPORASO, F. 1991. The Galápagos tortoise conservation program: the plight and future for the Pinzón Island tortoise. In: Beaman, K.R., Caporaso, F., McKeown, S, and Graff, M.D., eds., Proc. 1st Int. Symposium Turt Tort: Conservation and Captive Husbandry.

CASTRO, M. 1970. Chatham Island tortoise. Animals (London) 13(1):43-45.

CAYOT, L.J. 1985. Effects of El Niño on giant tortoises and their environment. In: G. Robinson and E. M. del Pino, eds., El Niño en las Islas Galápagos. El Evento de 1982-1983. Charles Darwin Foundation, Quito, Ecuador, pp. 363-398.

CAYOT, L.J. 1987. Ecology of giant tortoise (*Geochelone elephantopus*) in the Galápagos Islands. Ph.D. Thesis, Syracuse Univ., Syracuse, New York.

CAYOT, L.J., AND LOUIS, E. 1994. Recent increase in killing of giant tortoises on Isabela Island. Noticias de Galápagos 54:2-7.

CAYOT, L.J., AND MÁRQUEZ, C. 1994. Informacion sobre las poblaciones de Galápagos en Isabela. Carta Informativa, Estacion Cientifica Charles Darwin 38:3.

CAYOT, L.J., AND MORILLO, G.E. In press. Rearing and reintroduction of Galápagos tortoises: *Geochelone elephantopus hoodensis*, a case study. In: Conservation, restoration and management of tortoises and turtles.

CAYOT, L.J., MUÑOZ, E., AND MORILLO, G. 1993. Conservacion de las tortugas de la Isla Pinzón. Informe Anual de la Estacion Cientifica Charles Darwin 1988-89. Quito, Ecuador.

CAYOT, L.J., AND SNELL, H.M. 1996a. Tortoises from Cerro Paloma, Isabela threatened. Noticias de Galápagos 56:2.

CAYOT, L.J., AND SNELL, H.M. 1996b. Goats damage Volcán Alcedo, Isabela. Noticias de Galápagos 56:3.

COLNETT, J. 1798. Voyage to the South Atlantic and round Cape Horn into the Pacific Ocean, for the purpose of extending the spermaceti whale fisheries. London, 179 pp.

COOKSON, J. 1876. On the tortoises etc. of the Galápagos Islands. Proc. Zool. Soc. London 1876:178-179.

CORLEY SMITH, G.T. 1977. The present status of the giant tortoise *Geochelone elephantopus* on the Galápagos islands. Int. Zool. Yearbook 17:109-112.

CRUMLY, C.R. 1982. A cladistic study of *Geochelone* using cranial osteology. J. Herpetol. 16(3):215-234.

CRUMLY, C.R. 1984. The cranial morphometry of Galápagos tortoises. Proc. California Acad. Sci. 43(9):111-121.

CRUMLY, C.R. 1985. A hypothesis for the relationships of land tortoise genera (family Testudinidae). Studia Geologica Salmanticensia. Vol. Esp. 1. Studia Palaeocheloniologica I:115-124.

CRUZ, F., MONTESDEOCA, J., AND ZAMBRANO, R. 1994. El incendio forestal en la isla Isabela. Carta Informativa, Estacion Científica Charles Darwin 38:4-6.

CRUZ, M. 1994. The story of the discovery of the tortoise "Lonesome George" on Pinta island. Noticias de Galápagos 53:15-18.

CUVIER, G. 1829. La règne animal distribué d'après son organisation,... Nouvelle edition,... Tome II. Paris: Deterville, Crochard, 406 pp.

DAMPIER, W. 1697. A New Voyage Around the World. London.

DARWIN, C. 1839. Journal of Researches into the Geology and Natural History of the various countries visited by H.M.S. Beagle, under the command of Captain Fitz Roy, R.N., from 1832 to 1836. London: Henry Colburn, 637 pp.

DARWIN, C. 1845. Journal of Researches into the Natural History and Geology of the countries visited during the voyage of H.M.S. Beagle round the world, under the command of Capt. Fitz Roy, R.N. Second Edition, Corrected, with Additions. London: John Murray, 519 pp.

DELANO, A. 1817. A narrative of voyages and travels in the Northern and Southern hemispheres comprising three voyages around theworld together with a voyage of survey and discovery in the Pacific Ocean and Oriental Islands. Boston, pp. 369-385.

DE ROY MOORE, T. 1979. The browsing tortoise. Pacific Discovery 32(1):11-17.

DE SOLA, R. 1929. Notes on sex determination in a species of Galápagos tortoise. Copeia 1929:54.

DE SOLA, R. 1930. The Liebespiel of *Testudo vandenburghi*, a new name for the mid-Albemarle Island Galápagos tortoise. Copeia 1930:79-80.

DE VRIES, T. 1968. On the spot. Int. Turtle Tortoise Soc. J. 2(2):16-19,33.

DE VRIES, T. 1984. The giant tortoise: a natural history disturbed by man. In R. Perry, ed.: Key Environments: Galápagos. Pergamon Press, pp.145-156.

DORST, J. 1965. Découverte d'une population des tortues a San Cristóbal. Noticias de Galápagos 4:19.

DOWLING, H.G. 1962. World's end revisited. Animal Kingdom 65:130-142.

DUMÉRIL, A.M.C., AND BIBRON, G. 1835. Erpétologie Générale ou Histoire Naturelle des Reptiles. Tome 1er. Lib. Encyc. Roret, Paris.

EIBL-EIBESFELDT, I. 1959. Survey on the Galápagos Islands. UNESCO Mission Report 8:1-31.

EIBL-EIBESFELDT, I. 1960. Galápagos. London: MacGibbon and Kee.

ENRIQUEZ, R. 1988. Important aspects of the lifecycle of the Galápagos tortoise (*Geochelone elephantopus darwinii*) on Santiago Island. CDRS Annual Report, pp. 136-137.

ERNST, C., AND BARBOUR, R. 1989. Turtles of the World. Washington, Smithsonian Inst.

EVANS, L.T. 1949. The reproductive behavior of the giant tortoises, *T. vicina* and *T. vandenburghi*. Anat. Rec. 105(3):506.

EVANS, L.T., AND QUARANTA, J.V. 1949. Patterns of cooperative behavior in a herd of 14 giant tortoises at the Bronx Zoo. Anat. Rec. 105(3):506.

EVANS, L.T., AND QUARANTA, J.V. 1951. A study of the social behavior of a captive herd of giant tortoises. Zoologica 36(3):171-181.

FITZINGER, L. 1824. Sur des Reptiles observés a la Menagerie de Vienne. Bull. sci. nat. Paris 1:91-92.

FITZINGER, L. 1835. Entwurf einer systematischen Anordnung der Schildkröten nach den Gründsatzen der natürlichen Methode. Ann. Wien. Naturg. Mus. 1:105-128.

FITZINGER, L. 1843. Systema Reptilium. Fasciculus Primus, Amblyglossae. Vienna: Braunmuller et Seidel.

FITZ ROY, R. 1839. Narrative of the surveying voyage of His Majesty's ships Adventure and Beagle, between the years 1826 and 1836, describing their examination of the southern shore of South America, and the Beagle's circumnavigation of the globe. Vol. 2. Proceedings of the Second Expedition, 1831-36, under the command of Captain Robert Fitz-Roy, R.N. With appendix. London: Henry Colburn.

FOWLER, L. 1983. The population and feeding ecology of tortoises and feral burros on Volcan Alcedo, Galápagos islands. Ph.D. Thesis, Univ. of Florida, Gainesville.

FOWLER, L., AND JOHNSON, M.K. 1985. Diets of giant tortoises and feral burros on Volcan Alcedo, Galápagos. J. Wildl. Management 49(1):165-169.

FOWLER, L., AND ROE, L.H. 1984. Emergence success of tortoise nests and the effect of feral burros on nest success on Volcan Alcedo, Galápagos. Copeia 1984(3):702-707.

FRITTS, T.H. 1983. Morphometrics of Galápagos tortoises: evolutionary implications. In: Bowman, R.I. (Ed.). Patterns of Evolution in Galápagos organisms. Univ. California Press, Berkeley, pp. 107-122.

FRITTS, T.H. 1984. Evolutionary divergence of giant tortoises in Galápagos. Biol. J. Linn. Soc. 21(1-2):165-176.

FRITTS, T.H. 1988. Recognizing familiar faces. Noticias de Galápagos 46:11-12.

FRITTS, T.H., AND FRITTS, P.R. (Eds.). 1982. Race with extinction. Herpetological field notes of J. R. Slevin's journey to the Galápagos 1905-06. Herpetological Monograph 1.

GADOW, H. 1901. Amphibians and Reptiles. Cambridge Natural History. MacMillan and Co., London.

GARMAN, S. 1880 On certain species of Chelonioidae. Bull. Mus. Comp. Zoo. 6:123-126.

GARMAN, S. 1891. On a tortoise found in Florida and Cuba. Bull. Essex Inst. 23:141-144.

GARMAN, S. 1917. The Galapagos tortoises. Mem. Mus. Comp. Zool. 30(4):261-296.

GORDILLO, J., AND KENDRICK, A.W. 1989. Isla San Cristóbal. Noticias de Galápagos 48:21-22.

GRAY, J.E. 1825. A synopsis of the genera of reptiles and amphibians with a description of some new species. Ann. Phil. (2)10:193-217.

GRAY, J.E. 1831. Synopsis Reptilium, or short descriptions of the species of reptiles. Part 1. Cataphracta, Tortoises, Crocodiles and Enaliosaurians. London: Treuttel, Wurtz & Co.

GRAY, J.E. 1870. Notes on tortoises in the British Museum, with descriptions of some new species. Proc. Zool. Soc. Lond. 1870:653-659.

GRAY, J.E. 1873. Observation on chelonians, with descriptions of new genera and species. Ann. Mag. Nat. Hist. (4)11:289-308.

GROOMBRIDGE, B. 1982. The IUCN Amphibia-Reptilia Red Data Book. Part 1. Testudines, Crocodylia, Rhynchocephalia. IUCN, Gland, Switzerland.

GÜNTHER, A.C.L.G. 1875. Description of the living and extinct races of gigantic land-tortoises. Part 1-2. Introduction and the tortoises of the Galapagos Islands. Phil. Trans. Roy. Soc. London, Biol. Sci. 165:251-284.

GÜNTHER, A.C.L.G. 1877. The gigantic land tortoises (living and extinct) in the collection of the British Museum. Taylor and Francis, London.

GÜNTHER, A.C.L.G. 1896. Notes on *Testudo ephippium* Gthr. Novitates Zool. 3:329.

GÜNTHER, A.C.L.G. 1898. Recent progress on our knowledge of the distribution of the gigantic land tortoises. Proc. Linn. Soc. Lond. 110:14-29.

GÜNTHER, A.C.L.G. 1902. *Testudo galapagoensis* Baur. Described and figured by Günther. Novitates Zool. 9:184.

HALL, B. 1840. Extracts from a journal written on the coasts of Chili, Peru, and Mexico, in the years 1820-21-22. Edinburgh.

HARLAN, R. 1827. Description of a land tortoise, from the Galapagos Islands, commonly known as the "Elephant Tortoise." J. Acad. Natur. Sci. Philadelphia 5:284-292.

HAYES, F.E., AND BEAMAN, K. 1985. *Geochelone elephantopus vandenburghi* (Galápagos giant tortoise: Volcan Alcedo race). Morphology. Herp. Review 16(3):81-82.

HELLER, E. 1903. Papers from the Hopkins Stanford Galapagos Expedition, 1898-1899, XIV: Reptiles. Proc. Washington Acad. Sci. 5:39-98.

HENDRICKSON, J.R. 1965. Reptiles of the Galápagos. Pacific Discovery 18(5):28-36.

HENDRICKSON, J.R. 1966. The Galápagos tortoises, *Geochelone* Fitzinger (*Testudo* Linnaeus 1758 in part). In: Bowman, R.I. (Ed.). Proceedings of the Galápagos International Scientific Project. Univ. California Press, Berkeley, pp. 252-257.

HEYERDAHL, T. 1963. Archaeology in the Galápagos Islands. Occ. Pap. California Acad. Sci. 44:45-51.

HOECK, H. 1984. Introduced fauna. In: Perry, R. (Ed.). Key Environments. Galápagos. Pergamon Press, pp. 233-245.

HOLMAN, J.A. 1981. Florida sinkhole archaeological site. Expl. Jour. 69:114-116.

HONEGGER, R.E. 1972. Die Reptilien-Bestande auf den Galápagos-Inseln. Natur und Museum 102(12):437-447.

HONEGGER, R.E. 1980. Identification sheet for *Geochelone elephantopus*, produced for the Convention on International Trade in Endangered Species of Flora and Fauna [CITES]. Code A-301.011.003.005.

HOOIJER, D.A. 1971. A giant land tortoise, *Geochelone atlas* (Falconer and Cautley), from the Pleistocene of Timor. 1. Proc. Koninkl. Nederl. Akad. Wetench. B. 74:504-525.

ICZN (INTERNATIONAL COMMISSION ON ZOOLOGICAL NOMENCLATURE). 1985. International Code of Zoological Nomenclature. Third Edition. London: Int. Trust Zool. Nomencl., 338 pp.

IUCN. 1968. Red Data Book. Vilamil Mountain Tortoise *Testudo elephantopus guntheri*. Code number R/6/TESTU/ELE/GUN.

IUCN. 1975. Red Data Book. Vilamil Mountain Tortoise or South West Albemarle Tortoise. *Testudo elephantopus guentheri* Baur, 1889. Code 2.1.6.4.1.7 E.

IUCN. 1979. Red Data Book. List of threatened taxa. Code P.8.10.

IVERSON, J.B. 1985. Checklist of turtles of the world with English common names. SSAR Herp. Circular 14:1-14.

JIMENEZ FUENTES, E., AND CARBAJOSA TAMARGO, E . 1982. Un macho adulto de tortuga gigante del Miocene de Arevalo (Avila). Studia Geologica Salmanticensia 17:21-31.

JOHNSON, H. Undated, ca. 1912. Pioneers in Tropical America. London: Gresham Publ. Co.

JORDAN, K. 1896. On mechanical selection and other problems. Novit. Zool. 3:426-525.

KARL, S.A., BOWEN, B.W., AND AVISE, J.C. 1995. Hybridization among the ancient mariners: characterization of marine turtle hybrids with molecular genetic assays. J. Heredity 86:262-268.

LACÉPÈDE, B.G.E. 1788. Histoire naturelle des Quadrupèdes ovipares et des serpens. Vol. 1. Paris.

LATREILLE, P.A. 1801. Histoire naturelle des reptiles, Vol. 1.

LAZELL, J.D. 1989. Wildlife of the Florida Keys. A Natural History. Washington DC: Island Press, 253 pp.

LAZELL, J.D., AND MICHENER, M.C. 1976. This Broken Archipelago. Cape Cod and the Islands, Amphibians and Reptiles. New York: Demeter Press.

LEGUAT, F. 1708. Voyages et aventures de François Leguat, et de ses Compagnons, en deux Isles desertes des Indes Orientales. London: D. Mortier.

LEMONICK, M.D. 1995. Can the Galápagos survive? Time (30 Oct 1995):80-82.

LEVEQUE, R. 1963. La statut actuel des vertebres rare et menacés de l'Archipel des Galápagos. La Terre et la Vie 4:397-430.

LINDHOLM, W.A. 1929. Revidiertes Verzeichnis der Gattungen der rezenten Schildkröten nebst Notizen zur Nomenklatur einiger Arten. Zool. Anz. 81:275-295.

LOUIS, E. 1994. Galápagos tortoise mortality report for sites visited between July 13, 1994, and August 29, 1994. Unpublished report.

LOVERIDGE, A., AND WILLIAMS, E.E. 1957. Revision of the African tortoises and turtles of the suborder Cryptodira. Bull. Mus. Comp. Zool. 115:163-557.

MACFARLAND, C.G., AND REEDER, W.G. 1974. Cleaning symbiosis involving Galápagos tortoises and two species of Darwin's finches. Z. Tierpsychol. 34:464-483.

MACFARLAND, C.G., AND REEDER, W.G. 1975. Breeding, raising, and restocking of giant tortoises (*Geochelone elephantopus*) in the Galápagos Islands. In: Bowman, R.D. (Ed.). Breeding endangered species in captivity. Academic Press, London, pp. 13-37.

MACFARLAND, C.G., VILLA, J., AND TORO, B. 1974a. The Galápagos giant tortoises (*Geochelone elephantopus*). Part 1: status of surviving populations. Biol. Conserv. 6(2):118-133.

MACFARLAND, C.G., VILLA, J., AND TORO, B. 1974b. The Galápagos tortoises (*Geochelone elephantopus*). Part 2. Conservation methods. Biol. Conserv. 6(3):198-222.

MARLOW, R.W., AND PATTON, J.L. 1981. Biochemical relationships of the Galápagos giant tortoise (*Geochelone elephantopus*). J. Zool., London 195(3):413-422.

MÁRQUEZ, C. 1986. The giant tortoises and the great fire on Isabela. Noticias de Galápagos 44: 8.

MÁRQUEZ, C., GORDILLO, J., AND TUPIZA, A. 1994. The fire of 1994 and herpetofauna of southern Isabela. Noticias de Galápagos 54: 8-10.

MÁRQUEZ, C., MORILLO, G., AND CAYOT, L. 1991. A 25-year management program pays off: repatriated tortoises on Española reproduce. Noticias de Galápagos 51:17-18.

MÁRQUEZ, C., REA, S., FRITTS, T., CEPEDA, F., LLERENA, F., AND SNELL, H. 1992. Reproduccion de tortugas y iguanas terrestres en cautiverio en las Islas Galápagos. Acta Zool. Lilloana 41:295-308.

MAYR, E. 1963. Animal Species and Evolution. Harvard University Press.

MCBIRNEY, A.R., AND WILLIAMS, H. 1969. Geology and petrology of the Galápagos Islands. Geol. Soc. Amer. Memoir 118:1-197.

MERLEN, G. 1992. A dawn of despair: journey to Alcedo. Noticias de Galápagos 51:4-6.

MERLEN, G. 1995. Use and misuse of the seas around the Galápagos archipelago. Oryx 29(2):99-106.

MERREM, B. 1820. Versuch eines Systems der Amphibien. Marburg.

MERTENS, R., AND WERMUTH, H. 1955. Die rezenten Schildkröten, Krokodile und Brückenechsen. Zool. Jahrb. Abt. Syst. 83:323-440.

METZGER, S., AND MARLOW, R.W. 1986. The status of the Pinzón Island giant tortoise. Noticias de Galápagos 43:18-20.

MITCHILL, J. 1815. Description of the great Gallapago-Tortoise. From Mitchill's lectures on natural history. Medical Repository 2:309, 404.

MOORE, A., AND MOORE, T. 1982. Dog days on Isabela. Noticias de Galápagos 35:20-21.

MORILLO, G. 1992. Estado poblacional y ecologia reproductiva de la tortuga gigante de la isla Pinzón (*Geochelone elephantopus ephippium*), Galápagos. Tesis de Licenciatura, Pontifica Universidad Catolica del Ecuador, Quito.

MORILLO, G., AND CAYOT, L.J. 1990. Estado poblacional tortuga gigante de Galápagos de la Isla Pinzón. II Congreso Latinoamericano de Herpetologia. Mérida.

MUÑOZ, E. 1993. Ecology of feral goats (*Capra hircus*) on Alcedo Volcano. Noticias de Galápagos 52:2.

NORTH-COOMBES, A. 1979. The vindication of François Leguat. Service Bureau, Mauritius.

NOWAK, J.B., AND LAWESSON, J.E. 1988. Revegetation of the burnt area in Isabela. Noticias de Galápagos 46:18-19.

OBST, F.J. 1985. Die Welt der Schildkröten. Edition Leipzig.

PATTON, J.L., YANG, S.Y., AND MEYERS, P. 1975. Genetic and morphological divergence among introduced rat populations (*Rattus rattus*) of the Galápagos Archipelago, Ecuador. Syst. Zool. 24: 296-310.

PERRAULT, C. 1676. Suite des memoires pour servir a l'histoire naturelle des animaux. Paris, pp. 92-205.

PERRY, R.J. 1972. The Galápagos Islands. Dodd, Mead and Co.

PIÑOS O., F. 1987. Estudio sobre la sobreviviencia, el crecimiento, los movimientos y la alimentacion de los galapaguitos repatriados a Española. Technical report submitted to the Charles Darwin Research Station, Galápagos, Ecuador.

PORTER, D. 1815. Journal of a cruise made to the Pacific Ocean by Captain David Porter, in the United States Frigate Essex, in the years 1812, 1813, 1814. Philadelphia.

PRITCHARD, P.C.H. 1967. Living Turtles of the World. TFH Publ., Inc., New Jersey.

PRITCHARD, P.C.H. 1971a. A further report on Galapagos tortoises. Herpetol. Rev. 3(1):25.

PRITCHARD, P.C.H. 1971b. Galapagos tortoises, 1971. Herpetol. Rev. 3(3):49-51.

PRITCHARD, P.C.H. 1977. Three, two, one tortoise. Nat. Hist. 86(8):90-100.

PRITCHARD, P.C.H. 1979. Encyclopedia of Turtles. TFH Publ., Inc., Neptune, New Jersey.

PRITCHARD, P.C.H. 1984. Further thoughts on Lonesome George. Noticias de Galápagos 39:20-23.

PRITCHARD, P.C.H. 1985. The toughest tortoises in the Galapagos. Defenders 60(5):10-16.

PRITCHARD, P.C.H. 1986. A reinterpretation of *Testudo gigantea* Schweigger, 1812. J. Herpetol. 20(4):522-534.

PRITCHARD, P.C.H., AND TREBBAU, P. 1984. The Turtles of Venezuela. Contrib. in Herpetol. 2.

QUOY, J.R.C., AND GAIMARD, J.P. 1824a. Description d'une nouvelle espèce de tortue et de trois espèces nouvelles de Scinques. Bull. Sci. Nat. Geol., Paris: Zoologie, pp. 90-91.

QUOY, J.R.C., AND GAIMARD, J.P. 1824b. Sous-genre Tortue de Terre — *Testudo*. Brongn. Tortue Noire — *Testudo nigra*. N. In: Freycinet, M.L. de. Voyage autour du Monde executé sur l'Uranie et la Physicienne pendant les années 1817-1820, pp. 174-175.

REYNOLDS, R. 1982. Some observations on the captive management of Galápagos tortoises and land iguanas at the Darwin Station and suggestions for the future. 1982 Annual Report, Charles Darwin Research Station, Puerto Ayora, pp. 115-119.

REYNOLDS, T.N. 1835. Voyage of the United States Frigate Potomac under the Command of Commodore John Downes, During the Circumnavigation of the Globe, in the Years 1831, 1832, 1833, and 1834. New York.

RHODIN, A.G.J., PRITCHARD, P.C.H., AND REYNOLDS, R.P. 1983. Statusrapport för Galapagos jättesköldpaddor, *Geochelone elephantopus*, med anmärkningar om *G. e. ephippium* på Duncan. [Status report for Galapagos giant tortoises, *Geochelone elephantopus*, with observations on *G. e. ephippium* on Duncan.] Snoken - National Swedish Herpetological Association 13(3):67-72.

RODHOUSE, P., BARLING, R.W.A., CLARK, W.I.C., KINMOUTH, A.L., MARK, E.M., ROBERTS, D., ARMITAGE, L.E., AUSTIN, P.R., BALDWIN, S.P., BELLAIRS, A.D'A., AND NIGHTINGALE, P.J. 1975. The feeding and ranging behaviour of Galápagos giant tortoises (*Geochelone elephantopus*). The Cambridge and London University Galápagos Expeditions, 1972 and 1973. J. Zool., Lond. 176:297-310.

ROOSEVELT, K. 1930. The mountain party on Indefatigable Island. Bull. New York Zool. Soc. 33(4):156-163.

ROTHSCHILD, M. 1983. Dear Lord Rothschild. Glenside, PA: Balaban Publ., 398 pp.

ROTHSCHILD, W. 1896. Further notes on gigantic land tortoises. Novitates Zool. 3(2):85-91.

ROTHSCHILD, W. 1901. On a new land-tortoise from the Galapagos Islands. Novitates Zool. 8(4):372.

ROTHSCHILD, W. 1902a. On the habits and distribution of the Galapagos tortoises. Novitates Zool. 9:373.

ROTHSCHILD, W. 1902b. Note regarding *Testudo elephantopus*. Novitates Zool. 9:618.

ROTHSCHILD, W. 1902c. Further notes regarding *Testudo elephantopus*. Novitates Zool. 9:618.

ROTHSCHILD, W. 1902d. Description of a new species of gigantic land tortoise from the Galapagos Islands. Novitates Zool. 9:619.

ROTHSCHILD, W. 1903. Description of a new species of gigantic land tortoise from Indefatigable Island. Novitates Zool. 10:119.

ROTHSCHILD, W. 1915a. The giant land tortoises of the Galapagos Islands in the Tring Museum. Novitates Zool. 22:403-417.

ROTHSCHILD, W. 1915b. On the gigantic land tortoises of the Seychelles and Aldabra-Madagascar Group with some notes on certain of the Mascarene Group. Novitates Zool. 22:418-440.

ROTHSCHILD, W. 1925. Giant land tortoises of the Galapagos and Mascarene Islands. Proc. Linn. Soc. London 137:40.

ROTHSCHILD, W. 1928. Notes on gigantic land tortoises. Proc. Zool. Soc. London 1928:658-660.

RYDER, O.A. 1978. Laboratory tests confirm identification in giant tortoises. Zoonooz 51 (9): 13.

SCHAFER, S.F. 1982. The social behavior of the Galápagos tortoise, *Geochelone elephantopus*. M.S. Thesis, San Diego State University.

SCHAFER, S.F. 1983. Agonistic behavior of the Galápagos tortoise, *Geochelone elephantopus*, with emphasis on its relationship to saddle-backed shell shape. Herpetologica 39(4):448-456.

SCHNEIDER, J.G. 1783. Allgemeine Naturgeschichte der Schildkröten nebs einem systematischen Verzeichnisse der einzelnen Arten und zwey Kupfern. Leipzig.

SCHWARTZ, A., AND THOMAS, R. 1975. A check-list of West Indian Amphibians and Reptiles. Carnegie Mus. Nat. Hist., Spec. Publ. No. 1.

SCHWEIGGER, A.F. 1812. Prodromus monographiae Cheloniorum. Königsberg. Arch. Naturw. Math. 1: 271-368, 406-462.

SHAW, C.E. 1959. Four future giants. Zoonooz 32(12):3-5.

SHAW, C.E. 1961a. Breeding the Galapagos tortoise (*Geochelone elephantopus*). Int. Zoo Yearbook 3:102-104.

SHAW, C.E. 1961b. Another generation of Galapagos giants. Zoonooz 34 12):10-15.

SHURCLIFF, S.N. 1930. Jungle Islands: The "Illyria" in the South Seas. New York.

SIMKIN, T. 1984. Geology of Galápagos. In: Berry, R.J. (Ed.). Evolu-

tion in the Galápagos Islands. Biol. Jour. Linn. Soc. 21:61-75.
SLEVIN, J.R. 1931. Log of the schooner "Academy" on a voyage of scientific research to the Galápagos Islands, 1905-1906. Occ. Papers California Acad. Sci. 17:1-162.
SLEVIN, J.R. 1959. The Galápagos Islands. A history of their exploration. Occas. Pap. Calif. Acad. Sci. 25:1-150.
SMITH, H.M., AND SMITH, R.B. 1980. Synopsis of the herpetofauna of Mexico. VI. Guide to Mexican turtles. J. Johnson, Vermont.
SNOW, D.W. 1964. The giant tortoises of the Galapagos Islands. Oryx 7(6):277-290.
STAEDELI, J. 1958. New hope for giants of the past. Zoonooz 31(12):3-4.
STEADMAN, D.W. 1986. Holocene vertebrate fossils from Isla Floreana, Galápagos. Smithsonian Contrib. Zool. 413.
STEADMAN, D.W., AND ZOUSMER, S. 1988. Galápagos. Discovery on Darwin's Islands. Smithsonian Institution Press, Washington, D.C.
STRAUCH, A. 1862. Cheloniologische Studien mit besonderer Beziehung auf die Schildkrötensammlung der kaiserlichen Akademie der Wissenchaften zu St. Petersburg. Mem. Acad. Imper. Sci. St.-Petersbourg (7)5(7):1-196.
STRAUCH, A. 1865. Die Vertheilung der Schildkröten uber den Erdball. Ein zoogeographischer Versuch. Mem. Acad. Imp. Sci. St.-Petersbourg (7)8(13):1-207.
STRAUCH, A. 1890. Bemerkungen uber dis Schildkrötensammlung im zoologischen Museum der kaiserlichen Akademie der Wissenschaften zu St. Petersburg. Mem. Acad. Imper. Sci. St. Petersbourg (7)38(2):1-127.
STRAUCH, D. 1936. Satan Came to Eden. London: Jarrolds.
SULLOWAY, F.J. 1982. Darwin's conversion: the Beagle voyage and its aftermath. J. Hist. Biol. 15(3):325-396.
SULLOWAY, F.J. 1984. Darwin and the Galápagos. In: Berry, R.J. (Ed.). Evolution in the Galápagos Islands. Biol. J. Linn. Soc. 21(1-2):29-60.
SWITAK, K. 1995. Onan the terrible. Reptiles (Boulder) 3(8):48-55.
THORNTON, I. 1971. Darwin's Islands: a Natural History of the Galápagos. New York: Natural History Press.
THROP, J.L. 1969. Notes on breeding the Galapagos tortoise at the Honolulu Zoo. Int. Zool. Yearbook 9:30-31.
THROP, J.L. 1975. Notes on the management and reproduction of the Galápagos tortoise at the Honolulu Zoo. In: Martin, R.D. (Ed.). Breeding endangered species in captivity. Academic Press, London, pp. 39-42.
TIERNEY, J. 1985. Lonesome George of the Galápagos. Science 85(6):50-61.
TOWNSEND, C.H. 1924a. Impending extinction of the Galapagos tortoises. Bull. New York Zool. Soc. 27(2):55-56.
TOWNSEND, C.H. 1924b. New information on the Galapagos tortoises. Bull. New York Zool. Soc. 27(4):89-90.
TOWNSEND, C.H. 1925a. The whaler and the tortoise. Sci. Monthly 21:166-172.
TOWNSEND, C.H. 1925b. The Galapagos tortoises in their relation to the whaling industry. Zoologica 4(3):55-135.
TOWNSEND, C.H. 1928a. Propagation of the giant tortoise in the United States. Science 68(1750): 30.
TOWNSEND, C.H. 1928b. The Galapagos Islands revisited. Bull. New York Zool. Soc. 31(5):148-169.
TOWNSEND, C.H. 1930. The Astor Expedition to the Galapagos Islands. Bull. New York Zool. Soc. 33(4):135-155.
TOWNSEND, C.H. 1931a. Giant tortoises. Sci. Am. 144:42-44.
TOWNSEND, C.H. 1931b. Growth and aging in the giant tortoise of the Galapagos. Zoologica 9(13):459-474.
TOWNSEND, C.H. 1932. Galapagos tortoises in Bermuda. Bull. New York Zool. Soc. 35:121-123.
TOWNSEND, C.H. 1942. Can we save the giant tortoise? Bull. New York Zool. Soc. 55(5):109-111.
VAN DENBURGH, J. 1907. Preliminary description of four races of land tortoises from the Galapagos Islands. Proc. California Acad. Sci. (4)1:1-6.
VAN DENBURGH, J. 1914. The gigantic land tortoises of the Galapagos archipelago. Proc. Calif. Acad. Sci. (4)2(1):203-374.
VAN LIDTH DE JEUDE, T.W. 1898. On abnormal pectoral scutes in *Testudo ephippium* Gthr. Notes Leyden Mus. 20:126-128.
VANZOLINI, P.E. 1977. An annotated bibliography of the land and fresh-water reptiles of South America (1758-1975). Vol. 1 (1758-1900). Mus. Zool. Univ. Sao Paulo.
VILLA, J. 1972. Resumen de las actividades realizadas en la Isla Isabela en el presente periodo, 1971/1972. Volcan Sierra Negra y V. Cerro Azul. Manuscript.
VILLA, J., CHAPPY, O., AND TORO, B. 1971. Informe de las actividades realizadas en la programa de conservacion efectuado en la Isla Santiago (James) del 12 de Agosto al 22 de Diciembre. Manuscript.
WALBAUM, J.J. 1782. Cheloniographia oder Beschreibung einiger Schildkröten nach natürlichen Urbilden. Lubeck: J.F. Gleditsch.
WALLNER, L.M. 1995. Last of the giants. Reptiles. Guide to Keeping Reptiles and Amphibians 3(8):14-24.
WERMUTH, H., AND MERTENS, R. 1961. Schildkröten, Krokodile, Brückenechsen. Gustav Fischer Verlag, Jena.
WERMUTH, H., AND MERTENS, R. 1977. Liste der rezenten Amphibien und Reptilien. Testudines, Crocodylia, Rhynchocephalia. Das Tierreich 100:1-174.
WERNER, D. 1982. Social organization and ecology of land iguanas, *Conolophus subcristatus*, on Isla Fernandina, Galápagos. In: Burghart, G.M., and Rand, A.S. (Eds.). Iguanas of the World: their behavior, ecology, and conservation. New Jersey: Noyes Publ., pp. 342-365.
WHELAN, P.M., AND HAMANN, O. 1989. Vegetation regrowth on Isla Pinta: a success story. Noticias de Galápagos 48:11-13.
WIEGMANN, A.F.A. 1835. Bericht uber die Fortschritte der Zoologie in Jahre 1834. Amphibien. Arch. Naturg. 1(2):273-296.
WILLIAMS, E.E. 1950. *Testudo cubensis* and the evolution of western hemisphere tortoises. Bull. Amer. Mus. Nat. Hist. 95(1):1-36.
WILLIAMS, E.E. 1952. A new fossil tortoise from Mona Island, West Indies, and a tentative arrangement of the tortoises of the world. Bull. Amer. Mus. Nat. Hist. 99(9):541-560.
WILLIAMS, E.E. 1960. Two species of tortoises in northern South America. Breviora 120:1-13.
WILSON, E.O. 1994. Naturalist. Island Press/Shearwater Books. Washington D.C.
WILSON, E.O., AND BROWN, W.L. 1953. The subspecies concept and its taxonomic application. Systematic Zoology 2:97-111.
WITTMER, M. 1960. Floreana. London: M. Joseph.
WORAM, J.M. 1989. Galápagos island names. Noticias de Galápagos 48:22-32.

Received: 16 December 1992; *Revised*: 15 June 1995.
Accepted: 27 November 1995.